2015台达杯国际太阳能建筑设计竞赛获奖作品集
Awarded Works from International Solar Building Design Competition 2015

阳光与美丽乡村
Sunshine and Beautiful Village

中国可再生能源学会太阳能建筑专业委员会 编
Edited by Special Committee of Solar Buildings, CRES

执行主编 仲继寿 张磊
Chief Editor: Zhong Jishou, Zhang Lei

编辑 鞠晓磊 郑晶茹 夏晶晶
Editor: Ju Xiaolei, Zheng Jingru, Xia Jingjing

中国建筑工业出版社
CHINA ARCHITECTURE & BUILDING PRESS

图书在版编目(CIP)数据

2015台达杯国际太阳能建筑设计竞赛获奖作品集 阳光与美丽乡村/中国可再生能源学会太阳能建筑专业委员会编．—北京：中国建筑工业出版社，2015.8
 ISBN 978-7-112-18326-5

Ⅰ.①2… Ⅱ.①中… Ⅲ.①太阳能住宅-建筑设计-作品集-中国-现代 Ⅳ.①TU241.91-64

中国版本图书馆CIP数据核字（2015）第175895号

责任编辑：唐　旭　吴　绫
责任校对：李欣慰　关　健

2015台达杯国际太阳能建筑设计竞赛获奖作品集
阳光与美丽乡村
中国可再生能源学会太阳能建筑专业委员会　编
执行主编　仲继寿　张　磊
编辑　鞠晓磊　郑晶茹　夏晶晶
*
中国建筑工业出版社出版、发行（北京西郊百万庄）
各地新华书店、建筑书店经销
北京嘉泰利德公司制版
北京中科印刷有限公司印刷
*
开本：787×1092毫米　1/12　印张：23$\frac{1}{3}$　字数：450千字
2015年8月第一版　2015年8月第一次印刷
定价：**178.00元**（含光盘）
ISBN 978-7-112-18326-5
　　　　（27592）

版权所有　翻印必究
如有印装质量问题，可寄本社退换
（邮政编码 100037）

乡村孕育了城市，为城市提供劳动力、土地、产品、生态的保障。研发经济型宜居农村住房，让太阳能等清洁能源和绿色建筑技术应用到广大的新农村建设中，不仅是本次竞赛的关注点，更是今后城市反哺农村、促进人居可持续发展的重点。我们期待以竞赛为契机，和业内各界人士携手共建"美丽乡村"，实现"让城市融入大自然，让居民望得见山、看得见水、记得住乡愁"。

感谢台达环境与教育基金会资助举办2015台达杯国际太阳能建筑设计竞赛。

谨将本书献给致力于美丽乡村建设的设计者、建设者和践行者。

Villages are the lifeblood of cities; they provide cities with labor, land, products and ecological protection. An important aspect of this competition is to analyze the research done on economical livable rural housing, so that solar and other clean energy as well as green building technology can be applied to the majority of new rural construction; equally important, however, is the return of this clean energy and technology to rural areas and the promotion of sustainable development within rural settlements. We see this competition as an opportunity for industries from all walks of life to go hand in hand in building "Beautiful Villages"; to make sure we achieve our motto of "Integrating nature into cities, so that residents can see mountain-tops and water; and remember the pleasures of living in the countryside".

A warm thanks to the Delta Environmental & Educational Foundation for sponsoring the International Solar Building Design Competition 2015.

This book is dedicated to the designers, builders and practitioners of "Beautiful Village".

目 录
CONTENTS

阳光与美丽乡村　Sunshine and Beautiful Village

过程回顾　General Background

2015台达杯国际太阳能建筑设计竞赛评审专家介绍
Introduction of Jury Members of International Solar Building Design Competition 2015

获奖作品　Prize Awarded Works　001

综合奖·一等奖　General Prize Awarded · First Prize

风土再生（青海）　Regeneration of Vernacular Settlements (Qinghai)　002

日光·笙宅（黄石）　Solar Bamboo House (Huangshi)　008

综合奖·二等奖　General Prize Awarded · Second Prize

太阳礼赞　归宿阳光（青海）　Sun Praise Home of Sunshine (Qinghai)　014

长窠宅（青海）　Congregate Solar Housing (Qinghai)　018

光之结（青海）　Knot of Sunshine (Qinghai)　024

乐光·乐高·乐趣（黄石）　Sunshine · LEGO · Architecture (Huangshi)　030

综合奖·三等奖　General Prize Awarded · Third Prize

片山屋（青海）　Sliced Rockery House (Qinghai)　036

日月生辉——青海日月乡低能耗住房（青海）　The Brightness of Riyue—Low-energy Housing Project in Riyue of Qinghai (Qinghai)	042
庄窠·融光（青海）　Zero-carbon · Adaptable Sustainable Residential Design (Qinghai)	046
山间"屯"光（青海）　Indoor Sunshine and Breeze (Qinghai)	050
八分宅（黄石）　80% House (Huangshi)	054
墙·记忆（黄石）　Wall Memory (Huangshi)	060

综合奖·优秀奖　General Prize Awarded · Honorable Mention Prize

生态藏居（青海）　Ecological Tibetan Dwelling (Qinghai)	066
阳光庄窠（青海）　Hold Sunshine (Qinghai)	070
土生土长（青海）　Growth from the Loess (Qinghai)	074
阳光与美丽的交"措"（青海）　Sunshine Embrace Beauty (Qinghai)	080
光弧（青海）　Solar Arc (Qinghai)	086
"牧"光城——沐光·暮光·融光（青海）　The City of Pasture & Sunshine (Qinghai)	092
"1/2阳光"安多哇自宅（青海）　"A Half of Sunshine" AnDuoWa Home (Qinghai)	098
辞柯（青海）　CiKe (Qinghai)	104
光回故里（青海）　Sunning House (Qinghai)	110
日光宝盒（青海）　Sunshine in the Box (Qinghai)	116
土筑春意（青海）　Green Residence Built by Local Mud	122

一间阳光（青海）	Rooms of Sunshine (Qinghai)	128
双层聚落（青海）	Double-deck Settlement (Qinghai)	134
荒漠中的暖屋（青海）	Warm House (Qinghai)	138
乡村的守护与反哺（黄石）	Metabolism System of Village (Huangshi)	142
光井·模宅（黄石）	Solar Patio & Module House (Huangshi)	146
光临涵舍（黄石）	Light through Country House (Huangshi)	152
揽境（黄石）	Embrace the Environment (Huangshi)	158
乡人居所（黄石）	Villager Residence (Huangshi)	164
竹·光·园·间（黄石）	Bamboo · Light · Courtyard · Space (Huangshi)	170
家×宅（黄石）	Home × House (Huangshi)	174
沐浴阳光（黄石）	Green & Sunshine (Huangshi)	180
阳光加法（黄石）	Sunshine Plus (Huangshi)	184
"幸·盒·福"（黄石）	"Love · Box · Home" (Huangshi)	190
光映家箱（黄石）	Sunshine in the Box (Huangshi)	196
生·升不息（黄石）	Ecology Evolution (Huangshi)	202
材料之"煜"——乡村太阳能生态住宅设计（黄石）	The Glorious Sunshine of Materials—Rural Solar Ecological Residential Design (Huangshi)	208
集院宅（黄石）	Gardens in House (Huangshi)	214
耕·替（黄石）	Transforming (Huangshi)	220
"风""光"无限好——美丽家乡（黄石）	Sunshine · Wind · Endless (Huangshi)	226

有效作品参赛团队名单
Name List of all Participants Submitting Effective Works   232

2015台达杯国际太阳能建筑设计竞赛办法
Competition Brief for International Solar Building Design
Competition 2015   246

阳光与美丽乡村
Sunshine and Beautiful Village

2015年，正值国际太阳能建筑设计竞赛活动步入第十个年头，关注最具现实意义的热点问题，通过设计手段和工程实践传播低碳节能环保的生活理念是这项赛事不变的宗旨。本届竞赛聚焦新农村建筑，以"阳光与美丽乡村"为主题，设置农牧民定居青海低能耗住房项目和农村住房产业化黄石住宅公园项目两个赛题，面向全球征集作品，希望将太阳能等清洁能源和绿色建筑技术运用于广大农村建筑当中，推进中国新型城镇化、城乡一体化发展进程。

两个赛题均注重农民住房品质的提升，具有较强的现实意义。青海赛题强调农牧民定居及生态保护的问题；黄石赛题更突出工业化建造在推进新农村建设中可能发挥的作用。较之往届竞赛作品，本届作品的质量和数量都有很大提升。在关注绿色低碳、安全健康的设计要求的同时，充分考虑了太阳能利用技术对降低建筑使用能耗的作用，并在经济和技术层面更加突出实用与可操作性。特别是两个一等奖作品"风土再生"和"日光·笙宅"更是充分考虑了不同区域的气候特征和人们的生活方式，融入了当地的建筑文化，合理地采用了主被动太阳能技术，具有较强的可实施性。

让人备感欣喜的是，本届竞赛收到了一份来自中国人民大学附属中学高二年级同学提交的作品，她们是历届竞赛中最年轻的参赛者。虽然她们的作品未能进入终评，但年轻人广泛参与节能、环保建筑事业的行动和理念获得了评审专家的关注和肯定，组委会在终评会上特别安排了三名参赛者与国际评审专家面对面交流了她们的设计理念。

十年之间，竞赛影响力不断扩大，竞赛的平台效应日益凸显，正逐步成为行业智慧共享、新能源应用服务、获奖作品实践、创新人才培养和低碳理念传播的综合平台。本项赛事已经成为一个持续性的活动，先后有90余个参赛国家、5537个参赛团队参赛，提交了1032项有效作品。竞赛的参赛作品质量不断提高，并实现了竞赛之后的持续示范建造，"5·12汶川地震"后已经投入使用的杨家镇台达阳光小学、目前完成主体建设的中达低碳示范住宅，以及即将建设的青海农牧民定居农宅……这些项目正在发挥着良好的工程示范和理念传播作用。在总结和验证了多种实用、高效的创新技术方案的同时，竞赛培养了大批专业人才，不断为太阳能建筑事业注入新的力量。可以说，竞赛本身已经成为一个重要的绿色建筑行动。

感谢2015台达杯国际太阳能建筑设计竞赛的参与者，感谢所有关心与支持太阳能建筑发展的人们。

In 2015, the International Solar Building Design Competition celebrates its 10th anniversary. It maintains its ever unchanging purpose to focus on the hottest issues in modern architecture, spreading knowledge of design tools and engineering ideas for the practice of low-carbon green energy living. The focus of the competition is new rural construction, with "Sunshine and Beautiful Village" as the theme; The 2 competition topics are: Low Energy-Consumption Housing for Farmers in Qinghai Province; and Rural Housing Industrialization in the Residential Park, Huangshi City. With entries coming in from all around the world, it is our hope to integrate solar energy, clean energy and green building

农牧民定居青海低能耗住房项目场地实景图　Site of Low Energy-Consumption Housing for Farmers in Qinghai Province

technology into rural construction; which will promote China's urban-rural integration development process.

The 2 competition topics have strong practical significance and focuses on improving the quality of farmers' houses; The Qinghai theme emphasizes farmer settlements and ecological protection, while the Huangshi topic focuses on the role that industrialization plays in the construction of new rural areas. Compared to the work of previous competitions, the quality and quantity of this contest has seen great improvement. While focusing on low-carbon green building and safe design requirements, we also take into account solar energy technology and its role in reducing building energy use; all this is kept in mind as we analyze the practicality and feasibility in terms of economic and technical aspects. This is especially true when we look at the 2 top ranked works: "Regeneration of Vernacular Settlements" and "Solar Bamboo House"; They fully take into account the climatic characteristics of different regions and the way people live their lives. They've also assimilated with local building culture and reasonably use active and passive solar technology. The ability to implement these 2 works has a very strong chance of success.

It gives us great joy to announce that the current competition has an entry from a group of The High School Affiliated to Renmin University of China grade 2 students, making them the youngest group to ever participate in this competition. Although they were not able to enter the final round of assessment, the fact that these young adults were so enthusiastic and involved in energy-saving and environmental protection has caught the attention and recognition of our expert judges. During the final evaluation meeting, the organizing committee made special arrangements to set up a face-to-face between the 3 contestants and our international evaluation experts to discuss their design ideas.

Throughout the last decade, the influence of this competition has grown rapidly. It is now an integrated platform that shares industry wisdom, provides new energy services, finds new innovative talents and spreads the concept of low-carbon. This tournament has become an ongoing continuous event, with more than 90 participating countries, 5537 teams and 1032 submitted valid entries. The quality of works submitted by participants is getting better every year, and has achieved sustained demonstrations after the competition. After the "Wenchuan Earthquake of 5·12" these works were implemented into the Delta Sunshine Elementary School in Yangjia Town; up to now, they've completed the main construction of the Delta low-carbon demonstration project. Another example is the upcoming construction of Qinghai Low Energy-Consumption Housing for Farmers. These types of construction play a crucial role in the spread of demonstration projects and ideas. In the process of summarizing and verifying a large variety of practical, efficient and innovative technological solutions, the competition has cultivated many professionals, and continues to fill the solar building cause with strength and vitality. In other words, the competition itself has become a vital part of green building initiatives.

A heartfelt thanks to all the participants of the International Solar Building Design Competition 2015; and to all those out there who care about and support Solar Building Development.

农村住房产业化黄石住宅公园项目实景图　Site of Rural Housing Industrialization in the Residential Park, Huangshi City

过程回顾
General Background

本届竞赛由国际太阳能学会和中国可再生能源学会联合主办；国家住宅与居住环境工程技术研究中心、中国可再生能源学会太阳能建筑专业委员会承办；台达环境与教育基金会独家冠名。在相关单位的通力配合和社会各界的大力支持下，竞赛组委会于2014年1月成立，先后组织了竞赛启动、媒体宣传、校园巡讲、作品注册与提交、作品初评与终评、技术交流等一系列活动。这些活动得到了海内外业界人士的积极响应和参与。

一、竞赛筹备

开展美丽乡村建设，是推进生态文明和城乡一体化建设、全面建成小康社会的需要。

筹备之初，竞赛组委会将赛题锁定为美丽乡村农宅，并对竞赛实地建设场地进行了认真的考察。竞赛组委会得到了住房和城乡建设部建筑节能与科技司、青海省科学技术厅、湖北省黄石市房地产管理局等单位的大力支持，本届竞赛题目最终确定为"阳光与美丽乡村"，共设置两个赛题，包括农牧民定居青海低能耗住房项目和农村住房产业化黄石住宅公园项目。通过组织专家实地考察，确定了设计竞赛的场地建设条件，并编制了竞赛设计任务书。

农牧民定居青海低能耗住房项目实地考察　On site visit of Low Energy-Consumption Housing for Farmers in Qinghai Province

This competition is organized conjointly by the International Solar Energy Society and Chinese Renewable Energy Society (CRES). The Competition operators are the China National Engineering Research Center for Human Settlements and the Special Committee of Solar Buildings, CRES. It is sponsored solely by the Delta Environmental & Educational Foundation. With full cooperation of all the relevant organizations, the Organization Committee for this competition was set up in January of 2014; they then went on to organize competition start up, media campaigns, campus tours, entry registration and submission, preliminary and final evaluations, technical seminars, etc. These activities have received very positive responses and active participation from industry experts both at home and abroad.

1. Competition Preparation

In order to carry out Beautiful Village construction, we needed to promote ecological civilization, urban-rural integration and an all-around healthy society.

During preparation, the organizing committee locked in on the topic of Beautiful Village farm houses, and organized investigation of the competition's location. The competition organizing committee has received great support from: Department of Building Energy Saving and Science and Technology, Ministry of Housing and Urban-Rural Development of the People's Republic of China; Qinghai Science and Technology Department; Huangshi City Real Estate Authority; among other companies. In the end, we decided on "Sunshine and Beautiful Village" as this year's competition topic; including Low Energy-Consumption Housing for Farmers in Qinghai Province; and Rural Housing Industrialization in the Residential Park, Huangshi City. Through vigorous field investigation by our experts, the construction conditions were determined, and a competition design mission statement was compiled.

2. Competition Start-up

On May 13th, 2014, the International Solar Building Design Competition 2015 launched in Xining City, Qinghai Province. Among the prestigious guests in attendance at the competition launch ceremony were: Shi Dinghuan, Counselor of the State Council and Chairman of China Renewable Energy Society; Han Aixing, Deputy Director of Department of Building Energy Saving and

农村住房产业化黄石住宅公园项目实地考察 On site visit of Rural Housing Industrialization in the Residential Park, Huangshi City

竞赛启动仪式嘉宾留影 Photographing of honored guests at the competition launch ceremony

二、竞赛启动

2014年5月13日，2015台达杯国际太阳能建筑设计竞赛在青海省西宁启动。国务院参事、中国可再生能源学会理事长石定寰、住房和城乡建设部建筑节能与科学技术司副司长韩爱兴、台达集团创办人暨台达环境与教育基金会董事长郑崇华、中国可再生能源学会太阳能建筑专业委员会主任委员仲继寿、青海省科技厅副厅长周卫星、青海省住房和城乡建设厅党组书记贾应忠、青海省住房和城乡建设厅总工程师熊士泊、湖北省黄石市房地产管理局局长刘昌猛等嘉宾出席并参与了竞赛启动仪式，共同为"2015台达杯国际太阳能建筑设计竞赛"揭幕。

本届竞赛的两个题目在设置上各有侧重，中国青海低能耗农牧民定居项目希望通过太阳能建筑技术和绿色建筑技术的运用，设计出低成本、高性能，满足农牧民居住生活需求的健康性安全和低碳宜居的农村住宅，从而探索城乡统筹、资源集约、生态宜居、和谐发展的城镇发展模式；中国湖北农村低碳住宅产业化项目重点解决产业化构件的设计、施工和应用问题，展示先进适用的住宅设计理念

Science and Technology, Ministry of Housing and Urban-Rural Development of the People's Republic of China; Zheng Chonghua, Delta Group Founder and Chairman of the Delta Environmental and Education Foundation; Zhong Jishou, Director of the Special Committee of Solar Buildings, CRES; Zhou Weixing, Deputy Director of the Qinghai Science and Technology Department; Jia Yingzhong, Secretary of the Department of Housing and Urban-Rural Development of Qinghai Province; Xiong Shibo, Chief Engineer of the Department of Housing and Urban-Rural Development of Qinghai Province; Liu Changmeng, Huangshi City Real Estate Bureau; Together, we unveiled for the "International Solar Building Design 2015 Competition".

The 2 topics of this year's competition each focus on specific areas. China's Qinghai low energy farming residents project hopes to use solar energy building technology and green building technology to design low cost, high performance farming residents that meet the requirements of agricultural and pastoral lifestyles in these rural areas; all while maintaining health safety and low-carbon standards. This will allow exploration of urban-

和建筑技术，推广产业化住宅应用，从而提高生产效率和农宅质量，实现低冲击开发。竞赛组委会希望通过竞赛这一平台，努力实践太阳能利用等绿色、低碳、健康技术，研发经济型宜居农村住房。

三、校园巡讲

国际太阳能建筑设计竞赛巡讲是本项活动的重要组成部分，自启动以来，得到了清华大学、天津大学、东南大学、重庆大学、山东建筑大学等国内众多建筑院校的大力支持，逐渐成为一项具有影响力的校园公益活动，也吸引了大批富有激情与梦想的青年设计师积极参与竞赛。

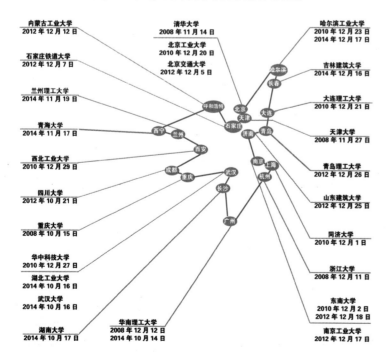

巡讲地图　Map of Campus Tours

rural planning, intensive resources, ecological livability and harmonious development of urban expansion models. China's Hubei rural low-carbon housing industrialization project focuses on solving the issues of design, construction and application of industrial components; it displays advanced concepts and construction technology for the application of residential designs, and promotes the industrialization of these applications; this in turn improves production efficiency and the quality of rural housing, and achieves low impact development. The Competition Organizing Committee hopes that through these platforms, we can strive to practice the use of solar energy through green, low-carbon and healthy technology; and from which we can research and develop economical and livable rural housing.

3. Campus Tours

Campus Tours for the International Solar Building Design Competition are an important part of this event. Since its inception, Tsinghua University, Tianjin University, Southeast University, Chongqing University, Shandong Jianzhu University and many other domestic architectural colleges and universities have given us their support; it has gradually become a very influential activity on campuses and has attracted a large number of passionate young designers to actively participate in the competition.

On October 14th, 2014, the "International Solar Building Design Competition 2015 Campus Tour" kicked off its first event at South China University of Technology; following this, it continued to spread its Solar Building design concepts at Hubei University of Technology, Wuhan University, Hunan University, Qinghai University, Lanzhou University of Technology, Jilin Jianzhu University, Harbin Institute of Technology, etc. The speaker on the tour was Zeng Yan, chief architect of the National Residential and Environmental Engineering Technology Research Center; the tour content covered the modern trends of application for solar building technology and analysis of works done by previous competition winners. Through this tour, students and teachers obtained an in-depth understanding of the Solar Building Design Competition and energy-saving technology; this in turn stimulated design inspiration within the competition's teams, and drastically increased registration for the competition.

华南理工大学巡讲现场　Campus Tour in South China University of Technology

武汉大学巡讲现场　Campus Tour in Wuhan University

湖北工业大学巡讲现场　Campus Tour in Hubei University of Technology

湖南大学巡讲现场　Campus Tour in Hunan University

青海大学巡讲现场　Campus Tour in Qinghai University

吉林建筑大学巡讲现场　Campus Tour in Jilin Jianzhu University

兰州理工大学巡讲现场　Campus Tour in Lanzhou University of Technology

哈尔滨工业大学巡讲现场　Campus Tour in Harbin Institute of Technology

2014年10月14日,"2015台达杯国际太阳能建筑设计竞赛校园巡讲"活动在华南理工大学首站拉开序幕,随后分别走进湖北工业大学、武汉大学、湖南大学、青海大学、兰州理工大学、吉林建筑大学和哈尔滨工业大学等院校传播太阳能建筑设计理念。巡讲主讲人为国家住宅与居住环境工程技术研究中心曾雁总建筑师,巡讲内容涵盖了太阳能建筑技术应用趋势和现状、历届竞赛获奖作品分析和本届竞赛介绍。通过巡讲,师生们对太阳能建筑设计竞赛和节能技术有了更深入的了解,激发了参赛团队的设计灵感,巡讲后注册人数明显上升。

四、媒体宣传

自竞赛启动伊始,组委会通过多渠道开展媒体宣传工作,包括:竞赛双语

4. Publicity through the Media

Since the start of the competition, the Organizing Committee propagated media through many channels, including: a bilingual website that reported real-time competition progress and popular science propaganda for solar buildings; set keyword searches on Baidu, making public searches much more convenient and much easier to log into the competition website; published special edition advertisements in China's "Architectural Journal", "Architecture Technique" and other professional magazines; published the competition's organization and advertisement situations in "Science and Technology Daily", "China Construction News", "Global Times" and more than 30 other printed media platforms; set up links and relevant information for the competition on Xinhua Net, Tencent, ABBS, Top

竞赛官方网站和宣传报道　Official website and media reports

网站实时报道竞赛进展情况并开展太阳能建筑的科普宣传；在百度设置关键字搜索，方便大众查询，从而更快捷地登陆竞赛网站。在中国《建筑学报》、《建筑技艺》等专业杂志刊登了竞赛活动宣传专版；在《科技日报》、《中国建设报》、《环球时报》等30余家平面媒体上发布了竞赛的组织与宣传情况；在新华网、腾讯网、ABBS、筑能网等50余家网站上报道或链接了竞赛的相关信息；同时，组委会与加拿大蒙特利尔大学、巴西麦肯锡教会大学、西班牙加泰罗尼亚理工大学等30余所国外院校取得联系并发布了竞赛信息。

五、竞赛注册及提交情况

本次竞赛的注册时间为2014年6月1日至2015年1月1日，共1232个团队通过竞赛官网进行了注册，其中，境外注册团队36个，包括日本、德国、法国、西班牙、加拿大、美国、瑞典、巴西、埃及、意大利、韩国等国家和中国港澳台地区。截至2015年3月1日，竞赛组委会收到德国、巴西、意大利、加拿大、埃及、中国香港和中国内地等国家和地区提交的参赛作品265个，其中有效作品250个。

六、作品初评

2015年3月5日，组委会将全部有效作品提交给初评专家组。每位专家根据竞赛办法中规定的评比标准对每一件作品进行评审，各自选出60份作品进入中评。经过竞赛评审专家的严格审查，3月10日组委会对所有专家的评审结果进行统计后，获得评审专家半数以上投票的共60份作品进入终评阶段。

七、作品终评

竞赛终评会于2015年4月8日在北京召开。经专家组讨论，一致推选M. Norbert Fisch教授担任本次终评工作的评审组长。在他的主持下，评审专家组按照简单多数的原则，集体讨论和公正客观地评选作品，通过四轮的投票，共评选出42项获奖作品，其中一等奖2名、二等奖4名、三等奖6名、优秀奖30名。

Energy and more than 50 other similar websites; at the same time, the Organizing Committee reached out to the University of Montreal, Universidade Presbiteriana Mackenzie, Universitat Polytechnic University of Catalonia in Spain, as well as more than 30 other foreign Universities to publish and promote the competition.

5. Registration and Works Submission

Registration for this competition was from June 1st, 2014 to January 1st, 2015; A total of 1232 teams registered via the competition's official website; among these, 36 teams were outside mainland China's borders, including Japan, Germany, France, Spain, Canada, USA, Sweden, Brazil, Egypt, Italy, Korea and China's Hong Kong, Macao and Taiwan regions. As of March 1st, 2015, the Organizing Committee has received 265 competition entries from Germany, Brazil, Italy, Canada, Egypt and Hong Kong and mainland China; among these, 250 were valid entries.

6. Preliminary Evaluation

On March 5th, 2015, the organizing committee submitted all works for preliminary evaluation. Each expert reviewed and assessed the criteria for the entries according to the rules of the competition, and selected 60 entries that they deemed worthy of further evaluation. On March 10th, after rigorous review from our competition experts, the Organizing Committee gathered all the data from the review and voted to select 60 entries for final evaluation (a majority vote was necessary).

7. Final Evaluation

The competition's final evaluation was held in Beijing on April 8th, 2015. During the final evaluation, Professor M. Norbert Fisch was unanimously elected as the leader for the final assessment review. Under his leadership, the expert review group collectively discussed and selected works objectively and fairly; using a simple principle of majority rule, they went through 4 rounds of voting and selected 42 winning entries; including 2 first place awards, 4 second place awards, 6 third place awards and 30 excellence awards.

终评会现场　Scenes of final evaluation conference

讨论作品　Discussion

终评专家组合影　Members of final evaluation jury

2015台达杯国际太阳能建筑设计竞赛评审专家介绍
Introduction of Jury Members of International Solar Building Design Competition 2015

评审专家
Jury Members

Deo Prasad：国际太阳能学会亚太区主席，澳大利亚新南威尔士大学建筑环境系教授

Mr. Deo Prasad, Asia-Pacific President of International Solar Energy Society(ISES) and Professor of Faculty of the Built Environment, University of New South Wales, Sydney, Australia

Peter Luscuere：荷兰代尔夫特大学 **(TU Delft)** 建筑系教授

Mr. Peter Luscuere, Professor of Department of Architecture, TU Delft, the Netherlands

M.Norbert Fisch：德国不伦瑞克理工大学教授（**TU Braunschweig**），建筑与太阳能技术学院院长

Mr. M.Norbert Fisch, Professor of TU Braunschweig, president of the Institute of Architecture and Solar Energy Technology, Germany

Mitsuhiro Udagawa：国际太阳能学会日本区主席，日本工学院大学建筑系教授

Mr. Mitsuhiro Udagawa, President of ISES-Japan and professor of Department of Architecture, Kogakuin University

崔愷：国际建协竞赛委员会委员、中国建筑学会常务理事、中国工程院院士、中国建筑设计院有限公司名誉院长、总建筑师

Mr. Cui Kai, Commissioner of International Union of Architects, Standing Director of Architectural Society of China, Academy of China Academy of Engineering, Honorary President and Chief Architect of China Architecture Design Group

喜文华：甘肃自然能源研究所所长，联合国工业发展组织国际太阳能技术促进转让中心主任，联合国可再生能源国际专家，国际协调员

Mr. Xi Wenhua, Director-General of Gansu Natural Energy Research Institute; Director-General of UNIDO International Solar Energy Center for Technology Promotion and Transfer; expert in sustainable energy field from United Nations, international coordinator

林宪德：台湾绿色建筑委员会主席，台湾成功大学建筑系教授

Mr. Lin Xiande, President of Taiwan Green Building Committee and Professor of Faculty of Architecture of Success University, Taiwan

冯雅：中国建筑西南设计研究院副总工程师，中国建筑学会建筑热工与节能专业委员会副主任

Mr. Feng Ya, deputy chief engineer of Southwest Architecture Design and Research Institute of China; deputy director of special committee of building thermal and energy efficiency, Architectural Society of China

仲继寿：中国可再生能源学会太阳能建筑专业委员会主任委员，国家住宅工程中心主任

Mr. Zhong Jishou, Chief Commissioner of Special Committee of Solar Building, Chinese Renewable Energy Society and Director of China National Engineering Research Center for Human Settlements

黄秋平：华东建筑设计研究院副总建筑师

Mr. Huang Qiuping, vice-chief architect of East China Architecture Design & Research Institute

获奖作品
Prize Awarded Works

风土再生 REGENERATION OF VERNACULAR SETTLEMENTS 01

综合奖·一等奖
General Prize Awarded · First Prize

注 册 号：3903
项目名称：风土再生（青海）
　　　　　Regeneration of Vernacular Settlements (Qinghai)
作　　者：王 鑫、梁 西、黄冠道、曹恺悦
参赛单位：北京交通大学

The annual average temperature: 3℃　　Highest temperature: 28.3℃
Minimum temperature: -23.5℃　　　　Year-round no cooling
Freezing line: 1.5 m　　　　　　　　The annual average sunshine hours: 2718.6 hours
Annual average wind speed: 1.73 m/s

Climate Analysis

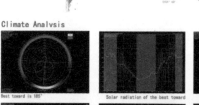

设计说明：

本设计以"风土再生"为主题，从人文生态角度，"风土"一是利用了当地的生土材料建造，解决冬季防风保温、夏季通风降温。再者借鉴并改善了传统庄窠的建筑形式，并加强和旧村落的联系，从单体建筑和群体环境都再次体现当地的风土人情。从气候地理环境，主要利用阳光中庭，并集主动式太阳能、沼气、地源热泵实现一体化供给能量并循环使用，同时从整体规划考虑太阳能发电的能源共享。从未来发展，为带动当地旅游产业，加入农家乐的居住形式，构建生态宜居、健康发展的新农村。

Design Notes:

With "Regeneration Vernacular Settlements" as the theme, the design from the perspective of human ecology, "terroir" First, the use of local materials to build adobe solve winter windproof insulation, ventilation and cooling in summer. Also learn and improve the profile of architectural forms of traditional village, and to strengthen ties and old villages, from single building and environmental groups are reflected in local customs again. From the climatic and geographical environment, the main advantage of the sun in the atrium, and set active solar, biogas, ground source heat pump to achieve the integration of the supply of energy and recycling, taking into account the share of energy from solar power overall planning. From future development, to promote the local tourism industry, adding farmhouse living forms, building livable, healthy development of the new countryside.

Han and Tibetan　　Agriculture and Livestock　　Religious holiday　　Now and old　　Comfort and tradition　　Tourism and New Countryside

专家点评：

作品院落空间组织合理，利用建筑群组及局部短墙形成西北封闭东南开放的空间围合，创造良好的院落风环境，平面布局紧凑、合理，整体造型有利于夏季通风和冬季防寒。充分结合青海高原地区冬季寒冷的气候特点和农牧民的生活方式，采用了现代主义建筑设计方法，利用当地具有民族特色的建筑材料，如厚重的夯土墙等，有利于保温和隔热。被动太阳能技术利用充分，阳光间及太阳能集热器布置合理。采用低成本技术策略，实现了低投入、高效率。建议适当放大中庭，效果会更好。

This entry uses rational courtyard space, and utilizes building groups and partial walls to form an enclosure with a closed off Northwest face and an open Southeast face; this gives a nice breeze to the courtyard environment. The layout is compact and reasonable; and the entire style is conducive to ventilation in the summer and protection against cold in the winter. It fully integrates the cold climate characteristics of the winter Qinghai Plateaus, and the lifestyles of local farmers and herdsmen. With a modernist approach to building design, it makes use of local ethnic building materials, such

风土再生 REGENERATION OF VERNACULAR SETTLEMENTS 02

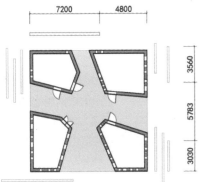

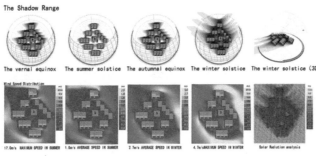

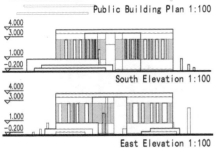

as compact earth walls, etc., which is a great insulator. It utilizes sufficient passive solar technology, a rational arrangement of solar room and collectors, and an overall technologically low-cost strategy to achieve a low-investment, high efficiency building. Enlargement of the atrium is suggested for a better effect.

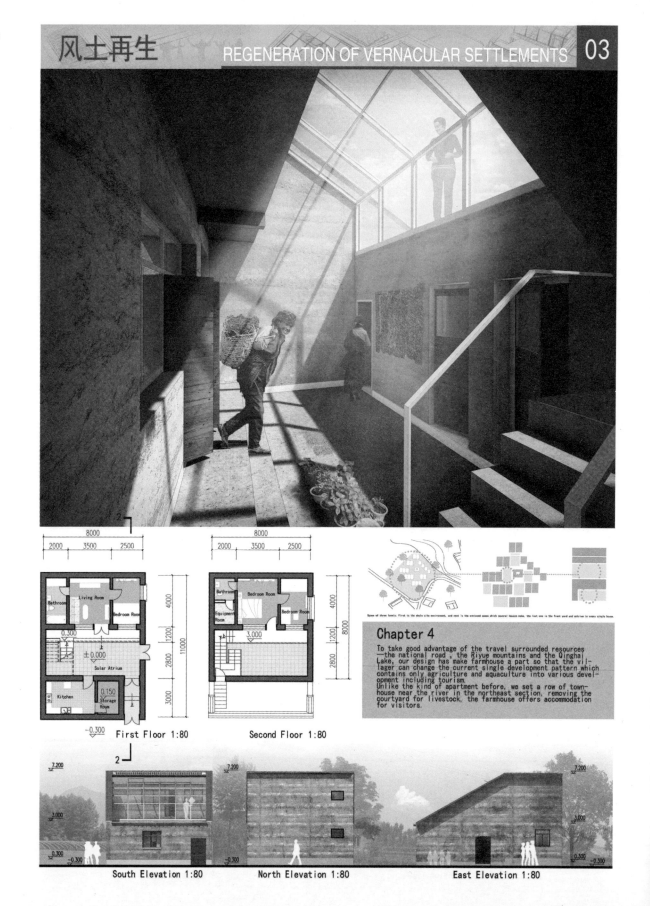

风土再生 REGENERATION OF VERNACULAR SETTLEMENTS 04

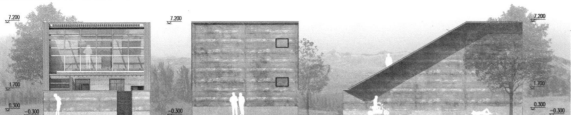

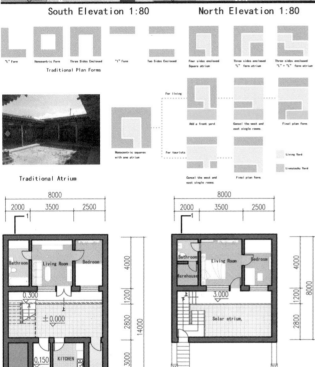

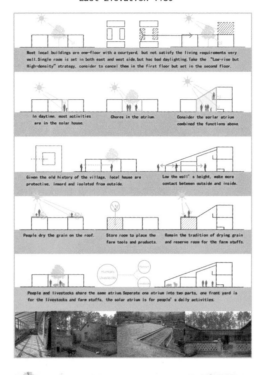

Chapter 4

Solar atrium. To change the original way that residents share the same yard with the livestock, we divided the only yard into two courtyards. The courtyard behind is the Solar atrium while the one before serve for the livestock. The solar atrium has greater interior space than the old heating sun room, front and rear of the house can be affected by sunlight into the atrium. Design makes more efficient use of the sun room focused.

Sunlight Analysis

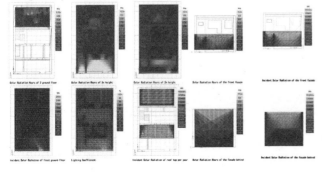

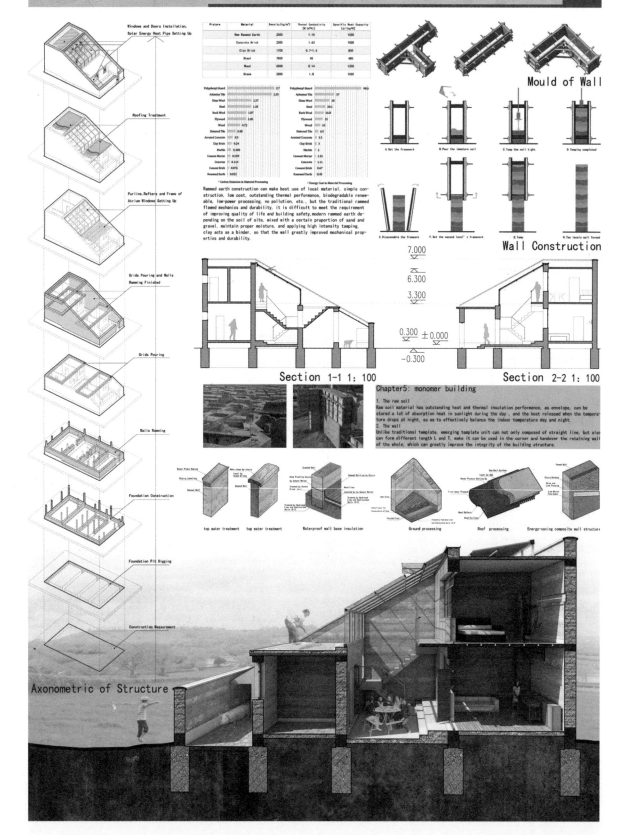

风土再生 REGENERATION OF VERNACULAR SETTLEMENTS 06

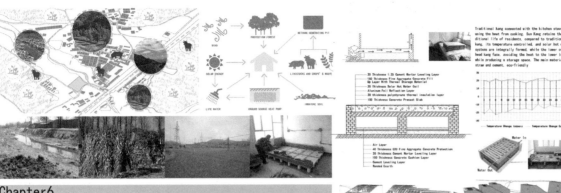

Chapter 6
1. The passive solar energy technology
In passive solar technology, uses the sun atrium improvement form, between the atrium with a layer of bedroom ground use pebble and generates heat. Glass atrium top use double deck glass so as to jointly maintain the winter construction internal temperature constant. Through the atrium of the internal thermal insulation shutter pull the glass door can be opened, and cooperate together to winter and summer seasons change, which arranged inside the room blinds on secondary regulation.

2. Active solar technologies
At the top of the atrium near the room can be arranged partially translucent yellow solar panels for easy piping layout. The set of household electrical energy delivered to the village level centralized grid common use. In the case without damaging the facade to 9 degrees split tube solar collectors placed on the roof, in the south facade of traditional houses rafters like to form a similar appearance. Collector tube hot water produced by combining ground source heat pump in the north side of the room and the overall circulation kang warm sun heating the room. The south side of the barn manure and kitchen use straw and other waste digesters daily treatment for the kitchen to provide clean energy for cooking, winter north of the collector tubes produced by the hot water and hot water kitchen food waste digesters keep the temperature constant.

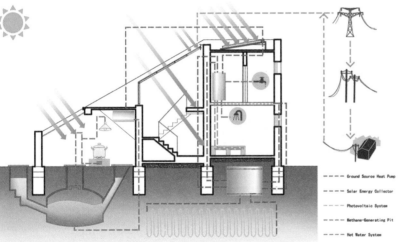

Wiring Diagrams

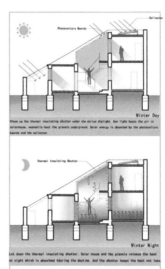

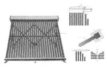

Photovoltaic Board | Ground Heat | Double Glazing | Glass Collector | Hot Water System

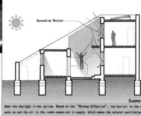

综合奖·一等奖
General Prize Awarded · First Prize

注 册 号：3301
项目名称：日光·笙宅（黄石）
　　　　　Solar Bamboo House
　　　　　(Huangshi)
作　者：刘红娟、许溪、吕怡卉、
　　　　虞鸿飞、白雪峰、黄琳茹
参赛单位：广西大学

专家点评：

作品探索了太阳能在热湿气候的长江流域的应用。建筑设计实用，并符合长江流域建筑特色。在被动太阳能利用方面，充分利用建筑体形实现了夏季建筑自遮阳和冬季太阳能利用；在主动太阳能利用方面，将太阳能热水系统、新风热回收系统等与建筑构造进行良好的结合；生物质能利用和雨水回收系统设计合理，实用性强。作品对装配式建筑和工业化建筑的发展具有清晰的理解，对模块的分解和组合进行了富有创新的探索，可实施性强。

This entry explores the application of solar energy in humid climate of the Yangtze River. The design is practical, and consistent with architectural features commonly used in the Yangtze River region. With respect to passive solar energy utilization, makes full use of the architectural shape in order to achieve self-shading and solar energy use in winter; with regards to active solar energy use, the building is in good conjunction with the solar hot water system and the new wind-heat recovery system. Biomass utilization and rainwater recycling system designs are reasonable and practical. This project has a clear understanding of the development

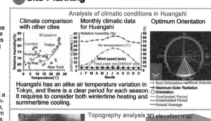
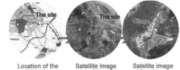

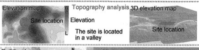

of assembly type buildings and industrial buildings; and approached the breakdown and buildup of this module with creativity; giving this entry strong implementation possibilities.

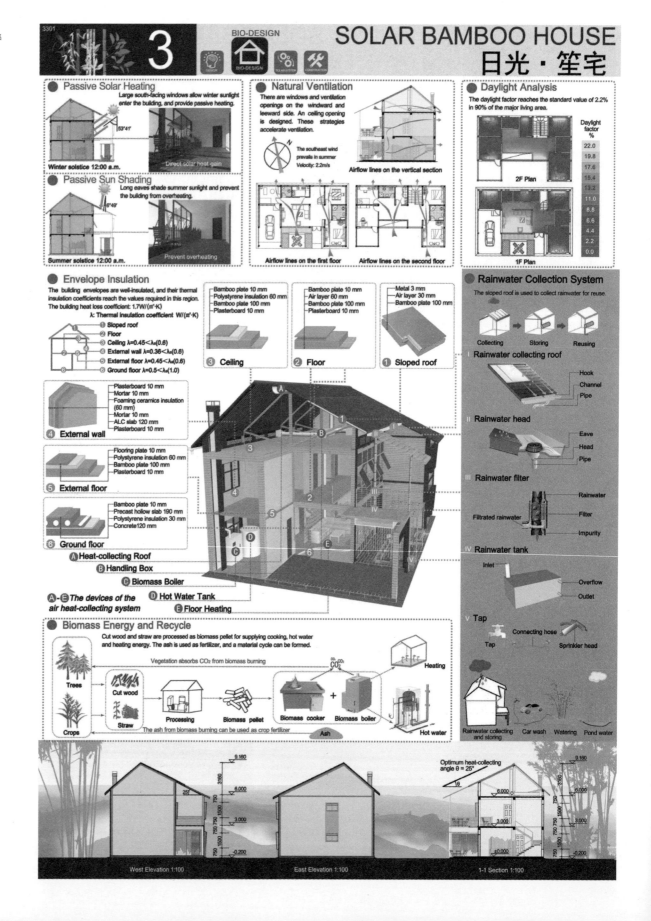

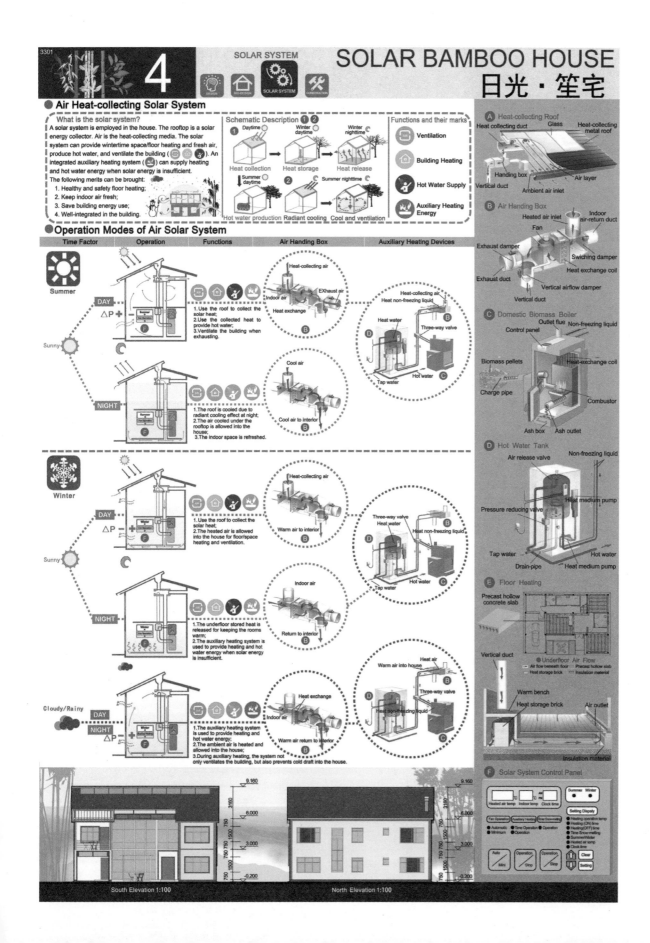

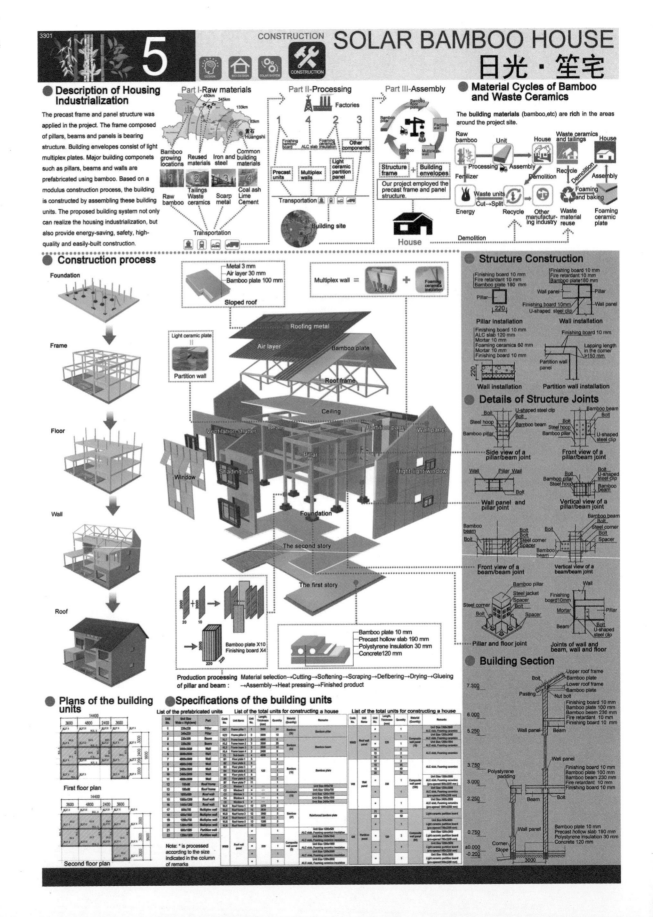

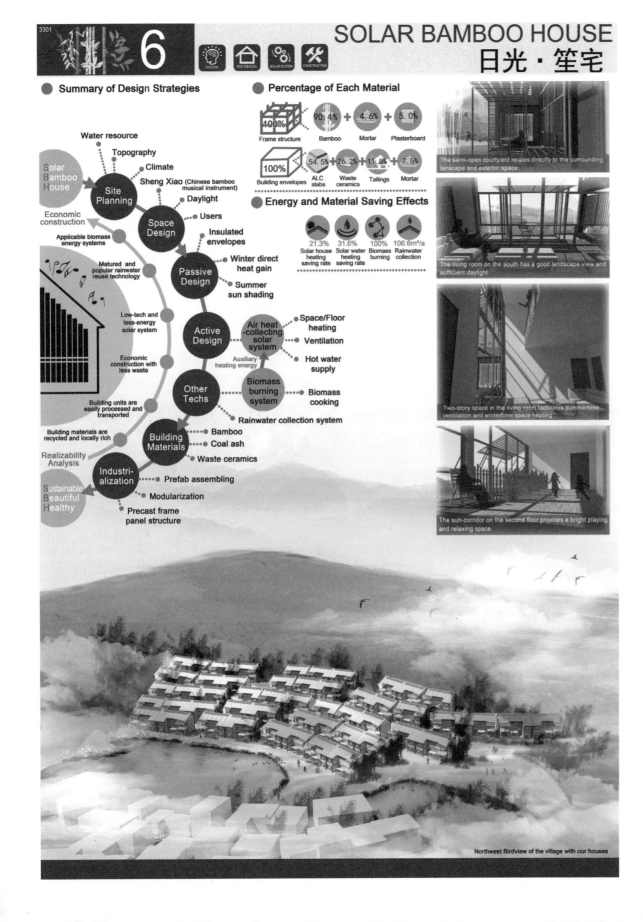

综合奖 · 二等奖
General Prize Awarded · Second Prize

注　册　号：3509
项目名称：太阳礼赞 归宿阳光（青海）
　　　　　Sun Praise Home of Sunshine (Qinghai)
作　　者：韩念森、张泽茜、王　惠、巴　婧
参赛单位：天津大学、山东建筑大学

专家点评：

作品规划布局富有特点，建筑空间变化丰富，具有流动性，建筑外形呈现出一种有趣的、具有地域传统的建筑风格。这种现代的藏式民居合理地利用了主被动的太阳能技术，集中布置的阳光走廊可以有效地降低成本，自然通风和室内热利用充分，火炕、沼气等传统技术利用得当，且可实施性强。作品没有体现出对建筑材料方面的考虑，太阳能与建筑结合设计欠缺。一体化程度不高。

This design is rich in variation and mobility. It presents a playful, yet distinctly traditional style architecture. These are clearly Tibetan houses, but with modern solar technology. The layouts are efficient, contributing to low cost solutions, with a centrally placed sunshine corridor. Solar energy is harnessed both passively and actively, and contributes to the energy efficiency of the houses. Traditional technologies like Kangs and Biogas are integrated in a logical way. This design does not reflect great consideration of construction materials, and the integration between the buildings and solar energy is not high.

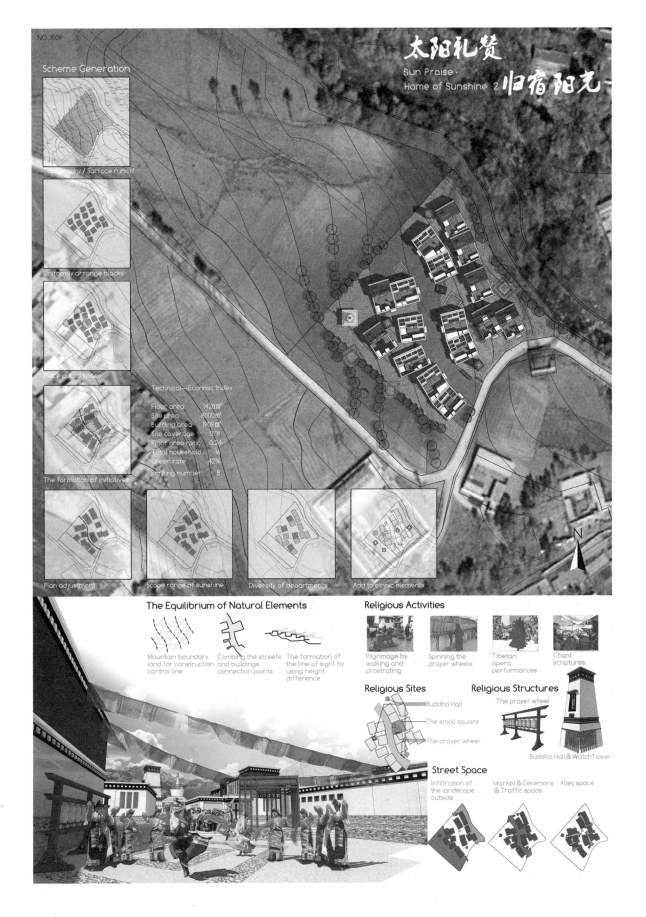

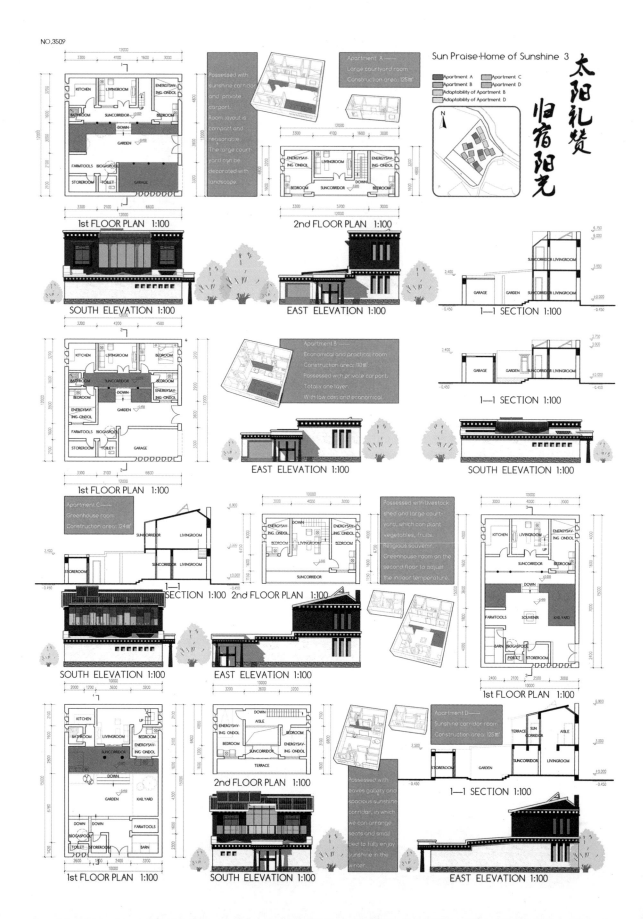

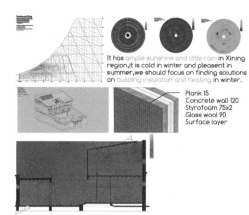

太阳礼赞 归宿阳光
Sun Praise·Home of Sunshine 4

It has ample sunshine and little rain in Xining region, it is cold in winter and pleasant in summer, we should focus on finding solutions on building insulation and heating in winter.

Plank 15
Concrete wall 120
Styrofoam 75×2
Glass wool 90
Surface layer

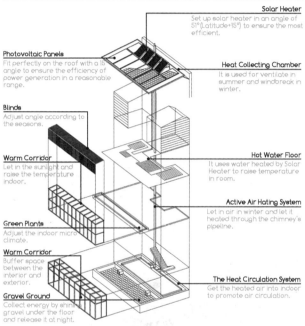

Solar Heater
Set up solar heater in an angle of 51°(Latitude+15°) to ensure the most efficient.

Photovoltaic Panels
Fit perfectly on the roof with a 15° angle to ensure the efficiency of power generation in a reasonable range.

Heat Collecting Chamber
It is used for ventilate in summer and windbreak in winter.

Blinds
Adjust angle according to the seasons.

Warm Corridor
Let in the sunlight and raise the temperature indoor.

Hot Water Floor
It uses water heated by Solar Heater to raise temperature in room.

Active Air Hating System
Let in air in winter and let it heated through the chimney's pipeline.

Green Plants
Adjust the indoor micro climate.

Warm Corridor
Buffer space between the interior and exterior.

The Heat Circulation System
Get the heated air into indoor to promote air circulation.

Gravel Ground
Collect energy by shining gravel under the floor and release it at night.

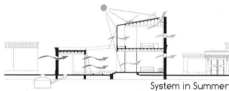

System in Summer
Close the blinds in summer, it will help reflect the sun, reduce thermal radiation, quantity of heat collected through the solar photovoltaic panel on the roof, providing electricity and hot water for users. Ventilate the air through windows in exterior walls and roof.

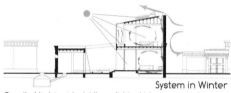

System in Winter
Open the blinds in winter, let the sunlight get into rooms, raise the temperature. Accessory warm corridor can reduce the heat loss of the opening part of the house. Heat regenerators under the floor, let them radiate quantity of heat to the interior.

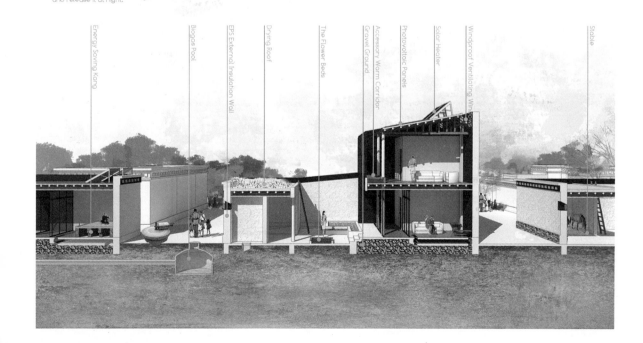

Energy Saving Kang | Biogas Pool | EPS External Insulation Wall | Drying Roof | The Flower Beds | Gravel Ground | Accessory Warm Corridor | Photovoltaic Panels | Solar Heater | Windproof Ventilating Windows | Stable

综合奖·二等奖
General Prize Awarded · Second Prize

注 册 号：3712
项目名称：长窠宅（青海）
　　　　　Congregate Solar Housing (Qinghai)
作　　者：钱世奇、沈宇驰、王建龙、刘洁莹
参赛单位：东南大学

专家点评：

设计以青海传统庄窠为原型，采用长院沿基地等高线联排规划出集中的建筑体量，不仅使每户的光照最佳，同时减少建筑能量流失。联排建筑群形成了抵御恶劣环境的屏障，同时将公共空间与农业暖棚包围在中心，形成了建筑群的太阳能应用体系。两进院落的单元组织形式，实现了功能合理分区，并使堂屋和卧室获得了最佳朝向。阳光间与特隆布墙能够在不同季节均对房屋的采暖通风进行有效调节；太阳能暖炕与相变天窗，既能够有效地节能又保留传统生活习惯；建筑外墙设置保温与抗震性能俱佳的复合夯土墙，既尊重材料特性，又体现当地建造特点。

This is a Qinghai traditional nest-type village design prototype; the rectangular courtyards are placed in a manner that follows the shape of the terrain; this not only provides the best household lighting, it also reduces building energy loss. The building groups act as a natural barrier to harsh environments; at the same time, public spaces and agricultural greenhouses are centered toward the middle, effectively forming the application of the solar energy system for the buildings. The courtyard is organized so that functions are

3712
CONGREGATE SOLAR HOUSING 长窠宅 ①

DESIGN CHALLENGE:
How to make the most use of solar energy by respecting the wisdom and culture of traditional technique and space of Zhuangke house?

设计挑战：
如何在继承保留传统庄窠宅的设计优势与文脉内涵的基础上，改良院落与公共空间的组织形式，实现对太阳能的更佳利用？

PROGRAM DERIVATION

Qinghai Traditional House | Design Prototype | East Room / West Room / South Room | Common Courtyard Wall | A Compound Cluster | Volume Dislocation to Echoing the Terrain

CONGREGATE SOLAR HOUSING
长窠宅 2

LOCATION

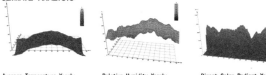

Huangyuan City → Xiangtuergan Town

SITE CONDITION RESEARCH
CLIMATE ANALYSIS

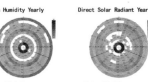

Average Temperature Yearly | Relative Humidity Yearly | Direct Solar Radiant Yearly
Wind Speed Yearly | Summer Wind Frequency | Winter Wind Frequency

YEAR-ROUND WEATHER CONDITION

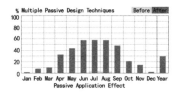

Project site is located in the eastern part of Qinghai Huangyuan. It is a continental monsoon climate, which has characteristics of short summer, long winter, the heating period up to 208 days, and windy throughout the year, less rainfall, long hours of sunshine (2718.6h) and solar radiation intensity.

DESIGN STRATEGY

1. Windproof and Heat Preservation in Heating Period, Insulation and Natural Ventilation in Cooling Period.
2. Passive Energy-saving Technology Mainly, Considering Positive Technology.
3. The Use of Solar Energy, Biomass and Other Renewable Energy Sources.
4. Local Material and Conventional Construction Method.

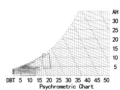

Psychrometric Chart

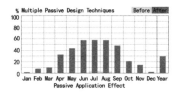

Multiple Passive Design Techniques — Passive Application Effect

South east of 15°
Optimum Orientation

COMPARISON OF ALTERNATIVE SCHEMES
SITE WIND SIMULATION

Scheme A | Scheme B | Scheme C | Scheme D

COOLING AND HEATING LOAD SIMULATION

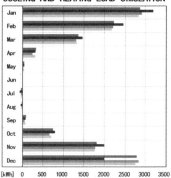

According to house size and site condition, 4 kinds of different plannings are developed and compared to each other.
Site wind simulation, cooling load and heating load are conducted.
As we can see, Scheme D is more effective in digesting northwest wind in winter, introducing northeast wind in summer. Meanwhile, Scheme D has less loads.

PLANNING STRATEGY

Best Orientation — Topography | Height Difference | Gentle Slope |

reasonably partitioned, with the main room and bedrooms facing the best direction. The solar room and Trombe wall are able to be adjusted effectively to ensure regulation of heating and ventilation during different seasons. Solar heated Kangs and phase changing sunroofs effectively save energy while retaining traditional village lifestyle habits; external walls are provided with insulation and seismic resistant compact earth walls; giving respect to material properties, while reflecting local characteristics.

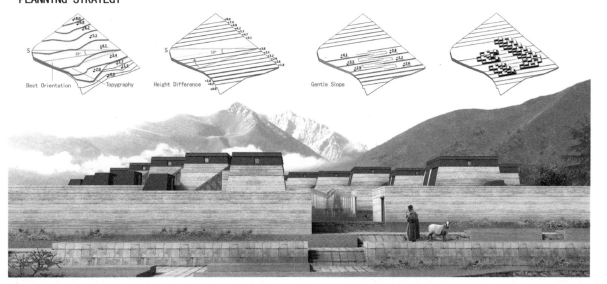

CONGREGATE SOLAR HOUSING

长橐宅 3

3712

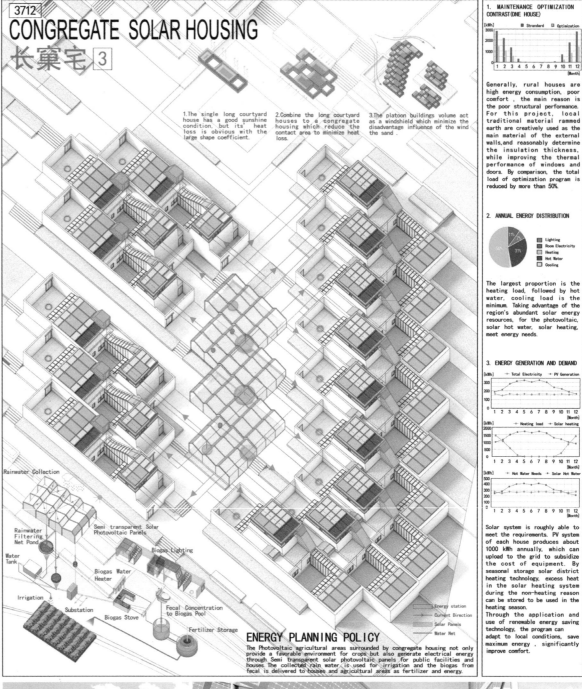

1. The single long courtyard house has a good sunshine condition, but its' heat loss is obvious with the large shape coefficient.

2. Combine the long courtyard houses to a congregate housing which reduce the contact area to minimize heat loss.

3. The platoon buildings volume act as a windshield which minimize the disadvantage influence of the wind the sand.

ENERGY PLANNING POLICY

The Photovoltaic agricultural areas surrounded by congregate housing not only provide a favorable environment for crops but also generate electrical energy through Semi transparent solar photovoltaic panels for public facilities and houses. The collected rain water is used for irrigation and the biogas from fecal is delivered to houses and agricultural areas as fertilizer and energy.

1. MAINTENANCE OPTIMIZATION CONTRAST (ONE HOUSE)

Generally, rural houses are high energy consumption, poor comfort, the main reason is the poor structural performance. For this project, local traditional material rammed earth are creatively used as the main material of the external walls, and reasonably determine the insulation thickness, while improving the thermal performance of windows and doors. By comparison, the total load of optimization program is reduced by more than 50%.

2. ANNUAL ENERGY DISTRIBUTION

The largest proportion is the heating load, followed by hot water, cooling load is the minimum. Taking advantage of the region's abundant solar energy resources, for the photovoltaic, solar hot water, solar heating, meet energy needs.

3. ENERGY GENERATION AND DEMAND

Solar system is roughly able to meet the requirements. PV system of each house produces about 1000 kWh annually, which can upload to the grid to subsidize the cost of equipment. By seasonal storage solar district heating technology, excess heat in the solar heating system during the non-heating reason can be stored to be used in the heating season. Through the application and use of renewable energy saving technology, the program can adapt to local conditions, save maximum energy, significantly improve comfort.

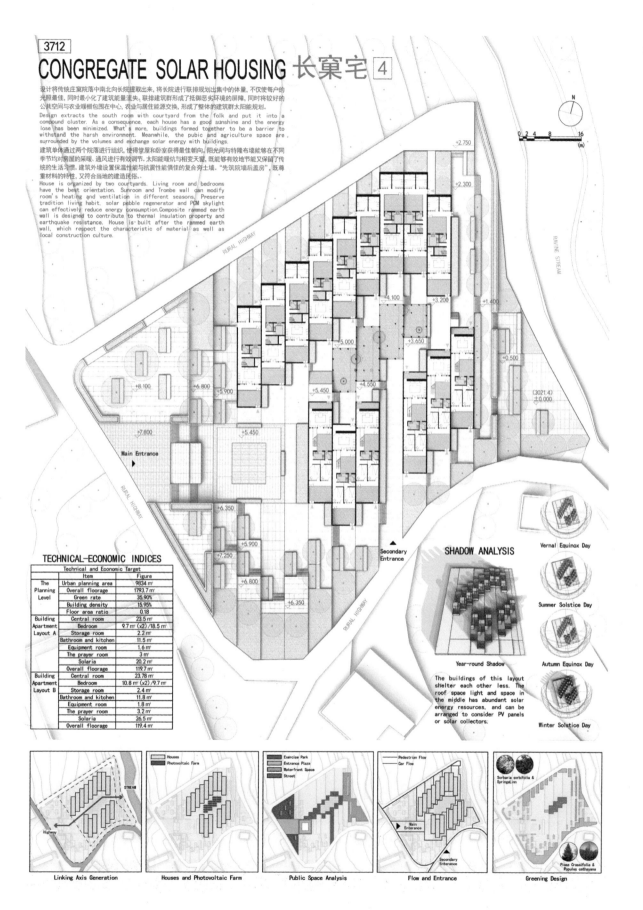

CONGREGATE SOLAR HOUSING 长窠宅 [4]

设计将传统庄窠落中南北向长院提取出来，将长院进行联排规划出集中的体量，不仅使每户的光照最佳，同时最小化了建筑能量直失。联排建筑群形成了抵御恶劣环境的屏障，同时将较好的公共空间与农业暖棚包围在中心，农业与居住能源交换，形成了整体的建筑群太阳能规划。

Design extracts the south room with courtyard from the folk and put it into a compound cluster. As a consequence, each house has a good sunshine and the energy lose has been minimized. What's more, buildings formed together to be a barrier to withstand the harsh environment. Meanwhile, the pubic and agriculture space are surrounded by the volumes and exchange solar energy with buildings.

建筑单体通过两个院落进行组织，使得堂屋和卧室获得最佳朝向。阳光间与特隆布墙能够在不同季节对房屋的采暖、通风进行有效调节。太阳能暖炕与相变天窗，既能够有效地节能又保留了传统的生活习惯。建筑外墙设置保温性能与抗震性能俱佳的复合夯土墙。"先筑院墙后盖房"，既尊重材料的特性，又符合当地的建造民俗。

House is organized by two courtyards. Living room and bedrooms have the best orientation. Sunroom and Trombe wall can modify room's heating and ventilation in different seasons. Preserve tradition living habit, solar pebble regenerator and PCM skylight can effectively reduce energy consumption.Composite rammed earth wall is designed to contribute to thermal insulation property and earthquake resistance. House is built after the rammed earth wall, which respect the characteristic of material as well as local construction culture.

TECHNICAL-ECONOMIC INDICES

	Technical and Economic Target	
	Item	Figure
The Planning Level	Urban planning area	9834 m²
	Overall floorage	1793.7 m²
	Green rate	35.90%
	Building density	15.95%
	Floor area ratio	0.18
Building Apartment Layout A	Central room	23.5 m²
	Bedroom	9.7 m² (x2) /18.5 m²
	Storage room	2.2 m²
	Bathroom and kitchen	11.5 m²
	Equipment room	1.6 m²
	The prayer room	3 m²
	Solaria	20.2 m²
	Overall floorage	119.7 m²
Building Apartment Layout B	Central room	23.78 m²
	Bedroom	10.8 m² (x2) /9.7 m²
	Storage room	2.4 m²
	Bathroom and kitchen	11.8 m²
	Equipment room	1.8 m²
	The prayer room	3.2 m²
	Solaria	26.5 m²
	Overall floorage	119.4 m²

SHADOW ANALYSIS

Year-round Shadow

The buildings of this layout shelter each other less. The roof space light and space in the middle has abundant solar energy resources, and can be arranged to consider PV panels or solar collectors.

Vernal Equinox Day
Summer Solstice Day
Autumn Equinox Day
Winter Solstice Day

Linking Axis Generation | Houses and Photovoltaic Farm | Public Space Analysis | Flow and Entrance | Greening Design

3712

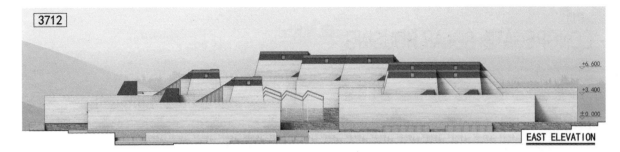

EAST ELEVATION

SUNROOM DESIGN DIAGRAM

Daylighting in Summer (A)

Daylighting in Summer (B)

Daylighting in Winter

Winter Night

Ventilation in Summer (A)

Ventilation in Summer (B)

Wind resistance in Winter (A)

Wind resistance in Winter (B)

Thermal Situation in Summer (A)

Thermal Situation in Summer (B)

Thermal Situation in Winter Day

Thermal Situation in Winter Night

PEBBLE REGENERATOR (A)

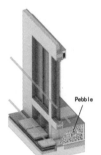
Winter Day

Winter Night

TROMBE WALL DIAGRAM

Glazing

Wood Frame

Black Painted Aluminum Sheet

Eps Insulation

Wood Frame

Section Detail 1:25

Winter

Summer and Transition Season

长寨宅 5

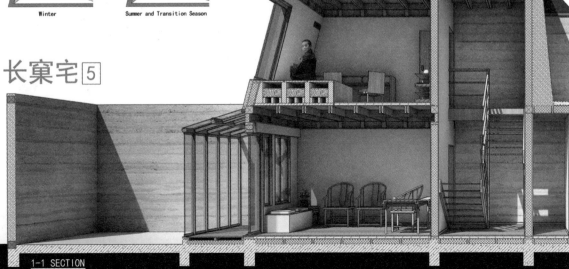

1-1 SECTION

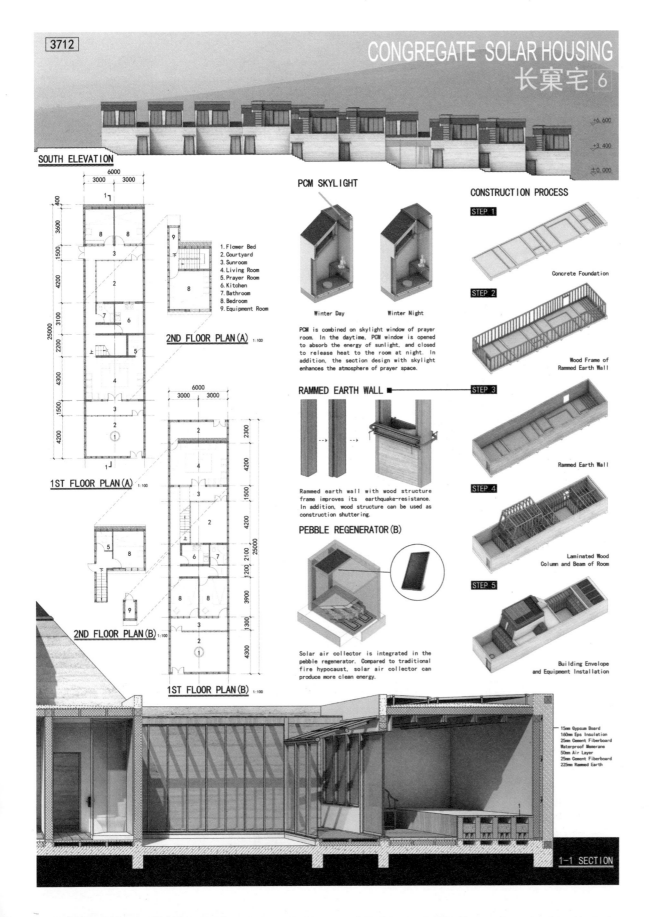

综合奖·二等奖
General Prize Awarded·
Second Prize

注 册 号：4141
项目名称：光之结（青海）
　　　　　Knot of Sunshine (Qinghai)
作　者：吴昌亮、孙 杰、徐 亮、
　　　　徐 斌、李哲健、刘大用、
　　　　马镇宇
参赛单位：东南大学、汉能全球光伏应用
　　　　　集团

专家点评：

作品采用吉祥结的图案，结合青海气候特点、藏区住宅建筑特色以及当地农民的生活习惯，形成了公共、半公共、私密层次递进的空间布局，营造了不同尺度的邻里交往空间，总体布局富有特点。单体设计通过四角通风塔、阳光房、追光百叶、节能保温墙体等设计，充分利用当地日照时间长、太阳辐射强的优势，解决冬季集热采暖和夏季通风降温的问题。建筑形象、色彩及局部图案具有青海藏式民居的特点。建议堂屋居中布置，外墙采用夯土等材料，更符合传统庄窠布局和当地建造方式。

This design uses an auspicious knot pattern; it combines Qinghai climate characteristics with Tibetan residential architecture, as well as the local lifestyles of farmers, to create progressive spatial layouts in public, semi-public and private level formations. This also provides different levels of space between neighbors, and an overall rich layout. Unit designs like ventilation towers, solar rooms, spotlight blinds, energy-saving insulation walls, etc., take advantage of the long hours of sunshine and strong solar radiation to solve the issue of heating

4141
LOCAL NATURAL LANDSCAPE

GEOGRAPHY ANALYSIS

XINING　　　　SITE

KNOT 光之结
OF SUNSHINE

LOCAL ARCHITECTURAL FEATURES

◇◇ SITE INVESTIGATION

The case on the basis of analysis of the present conditions of the site and the climatic feature of Qinghai, and the residential architectural features of Tibetan and habits of local farmers, simplifies the design of the Qinghai auspicious knot as the starting point for the design; connects the mountains and river by adjusting the layout of two residential groups; solves the problems of heating in winter and ventilation in summer with the ventilation towers, solar house, chasing light blinds, energy-saving insulation wall; and finally determines the placement of solar hot water and solar photovoltaic on the basis of the analysis of the light environment simulation. The meaning of the name "Knot of sunshine" characterizes the Tibetan auspicious knot, connects the mountains and river with residence, forms a harmonious neighborhood.

本方案在充分分析场地现状、青海的气候特点、藏式住宅的建筑特色以及当地农民的生活习惯的基础之上，简化了青海的吉祥结的图案意象作为设计的出发点；调整两组住宅组团的布局，形成了连接山与水的顺应地形的视廊，也营造了不同尺度的户间交流区域；通过四角的通风塔、阳光百房、追光百叶、节能保温墙体，充分利用当地日照时间长、太阳辐射强的优势，解决冬季集热采暖和夏季通风降温的问题；最后基于光环境分析确定太阳能集热器和太阳能光伏发热安放。
以"光之结"为名，一是表征藏式吉祥结的意象；二是指住区将山与水扣合联系；三是邻里交融和睦，通过咬合的住宅和多样的交流空间团结在一起。

LOCAL CLIMATE ANALYSIS

XINING which is located in the inland is a continental monsoon climate, with long hours of sunshine, strong solar radiation, dry and windy spring. There summer is short and cool, autumn is wet and rainy and winter is long and drying.

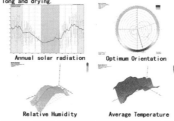

INSOLATION ANALYSIS

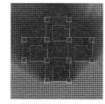 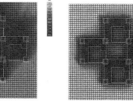 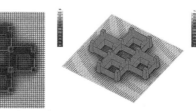

Big Chill Day (December 25th)　　The summer solstice (June 22nd)

GREEN DESIGN STRATEGIES

The more appropriate strategies for the project is utilization of solar energy, the use of high-performance building envelope materials, natural ventilation and lighting.

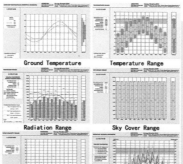

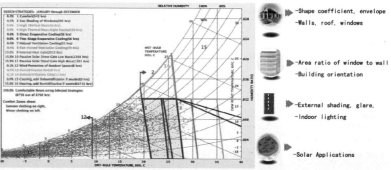

- Shape coefficient, envelope
- Walls, roof, windows
- Area ratio of window to wall
- Building orientation
- External shading, glare,
- Indoor lighting
- Solar Applications

CONCEPT

The design concept derives from the "auspicious knot" which can represents the Tibetan culture.

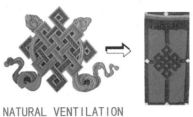

NATURAL VENTILATION

On the one hand, the exterior wall can stop the local cold monsoon; On the other hand, it can also form a negative pressure inside the courtyard, leading to the interior natural ventilation.

Tibet's traditional houses   Ventilation schematic

THERMAL ANALYSIS

In Summer Day:
vents in top and bottom open, air in solar house rises and discharged from vents at the top to remove excess heat. Meanwhile closed blinds block excessive solar radiation.

In Winter Day:
solar radiation warms the solar house up rapidly a part of the heat is stored in the sun roof/floor by the PCM heat storage material, the other warms the room to improve indoor thermal environment.

Night in Summer:
Height difference between the inlet and exhaust ports forms good indoor ventilation and comfortable thermal environment.

Night in winter:
stored heat in PCM heat storage material is released slowly to keep the temperature of the interior space, the double glazing can effectively prevent heat loss.

KNOT OF SUNSHINE
光之结

in winter and ventilation in the summer. Building images, color and patterns all have Qinghai Tibetan local residential characteristics. We recommend moving the rooms towards the middle, and to use compact earth wall materials, which is more in line with traditional village layouts and local construction methods.

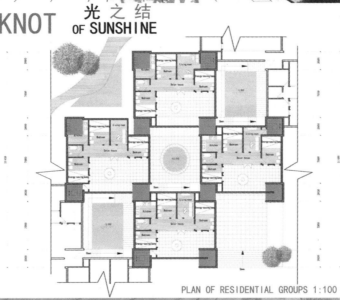

PLAN OF RESIDENTIAL GROUPS 1:100

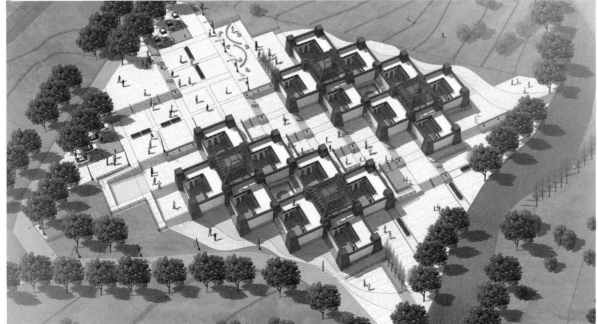

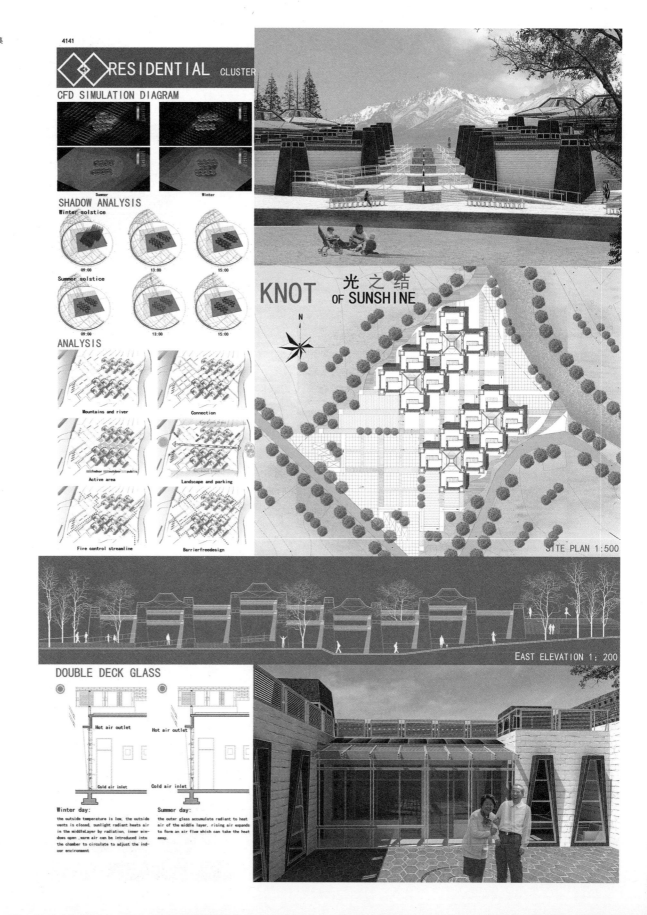

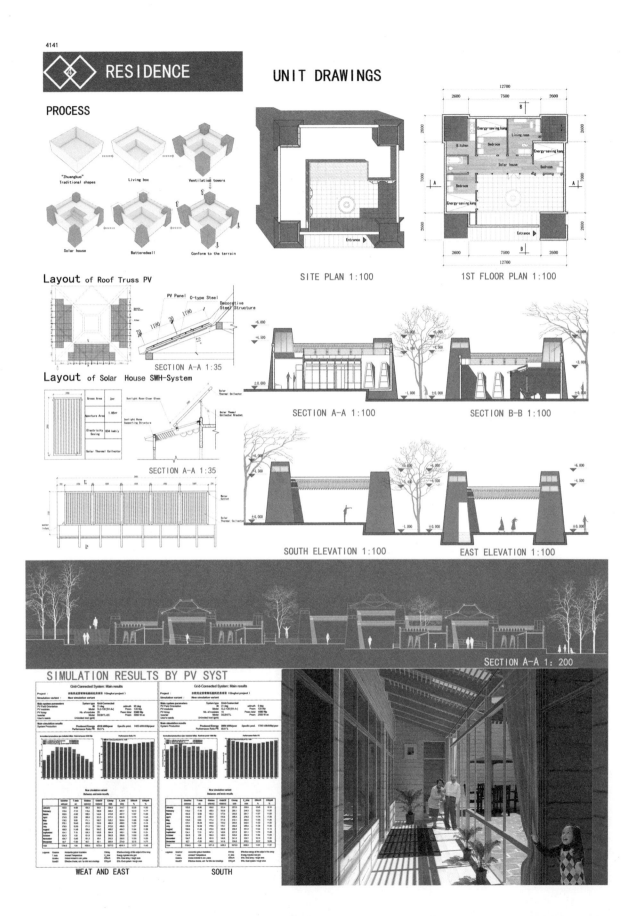

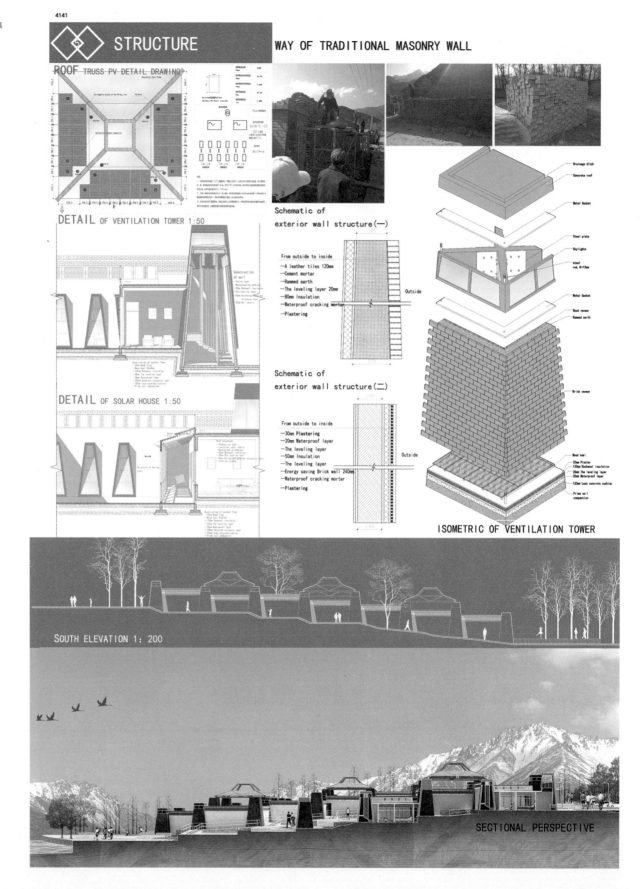

TECHNIQUE

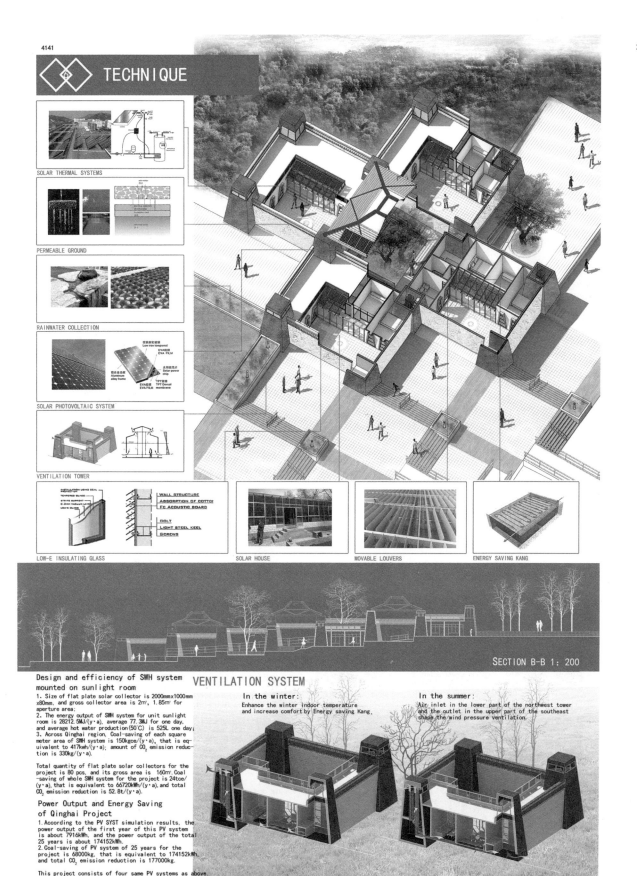

SOLAR THERMAL SYSTEMS

PERMEABLE GROUND

RAINWATER COLLECTION

SOLAR PHOTOVOLTAIC SYSTEM

VENTILATION TOWER

LOW-E INSULATING GLASS | SOLAR HOUSE | MOVABLE LOUVERS | ENERGY SAVING KANG

SECTION B-B 1:200

Design and efficiency of SWH system mounted on sunlight room

1. Size of flat plate solar collector is 2000mm×1000mm ×80mm, and gross collector area is 2m², 1.85m² for aperture area;
2. The energy output of SWH system for unit sunlight room is 28212.5MJ/(y·a), average 77.3MJ for one day, and average hot water production (50℃) is 525L one day;
3. Across Qinghai region, Coal-saving of each square meter area of SWH system is 150kgce/(y·a), that is equivalent to 417kwh/(y·a); amount of CO_2 emission reduction is 330kg/(y·a).

Total quantity of flat plate solar collectors for the project is 80 pcs, and its gross area is 160m². Coal-saving of whole SWH system for the project is 24tce/(y·a), that is equivalent to 66720kWh/(y·a), and total CO_2 emission reduction is 52.8t/(y·a).

Power Output and Energy Saving of Qinghai Project

1. According to the PV SYST simulation results, the power output of the first year of this PV system is about 7916kWh, and the power output of the total 25 years is about 174152kWh.
2. Coal-saving of PV system of 25 years for the project is 68000kg, that is equivalent to 174152kWh, and total CO_2 emission reduction is 177000kg.

This project consists of four same PV systems as above.

VENTILATION SYSTEM

In the winter:
Enhance the winter indoor temperature and increase comfort by Energy saving Kang.

In the summer:
Air inlet in the lower part of the northwest tower and the outlet in the upper part of the southeast shape the wind pressure ventilation.

综合奖·二等奖
General Prize Awarded · Second Prize

注 册 号：4265
项目名称：乐光·乐高·乐趣（黄石）
　　　　　Sunshine · LEGO · Architecture
　　　　　(Huangshi)
作　　者：陈慧贤、朱　婷、邵继中
参赛单位：南京工业大学

专家点评：

作品将工业化的概念贯穿设计始终，利用模块化理念，像乐高积木一样，通过卫生间模块、厨房模块、交通模块等组合形成多种户型；同时将采暖系统、供电系统和排水系统以及主动和被动式太阳能系统与建筑模块进行一体化结合，形成可在工厂预制的模块化产品，从而实现建筑的快速建造，提高建筑施工质量。作品在建筑造型表现上略显不足，缺少模块多样化组合形式的表达。

This design has taken the concept of Industrialization to the extremes. It consists of a modular set of LEGO-inspired building blocks, which when combined, form a great variety of individually composed houses; it utilizes bathroom modules, kitchen modules and transportation modules, among others, to create all kinds of different designs. At the same time, it combines heating, electricity, drainage, active solar energy and passive solar energy systems all into one; this way, the modules can be mass produced in factories, increasing the speed and quality of construction greatly. However, the architectural style is a little lacking, with little expression of the arrangement between modularity and diversification.

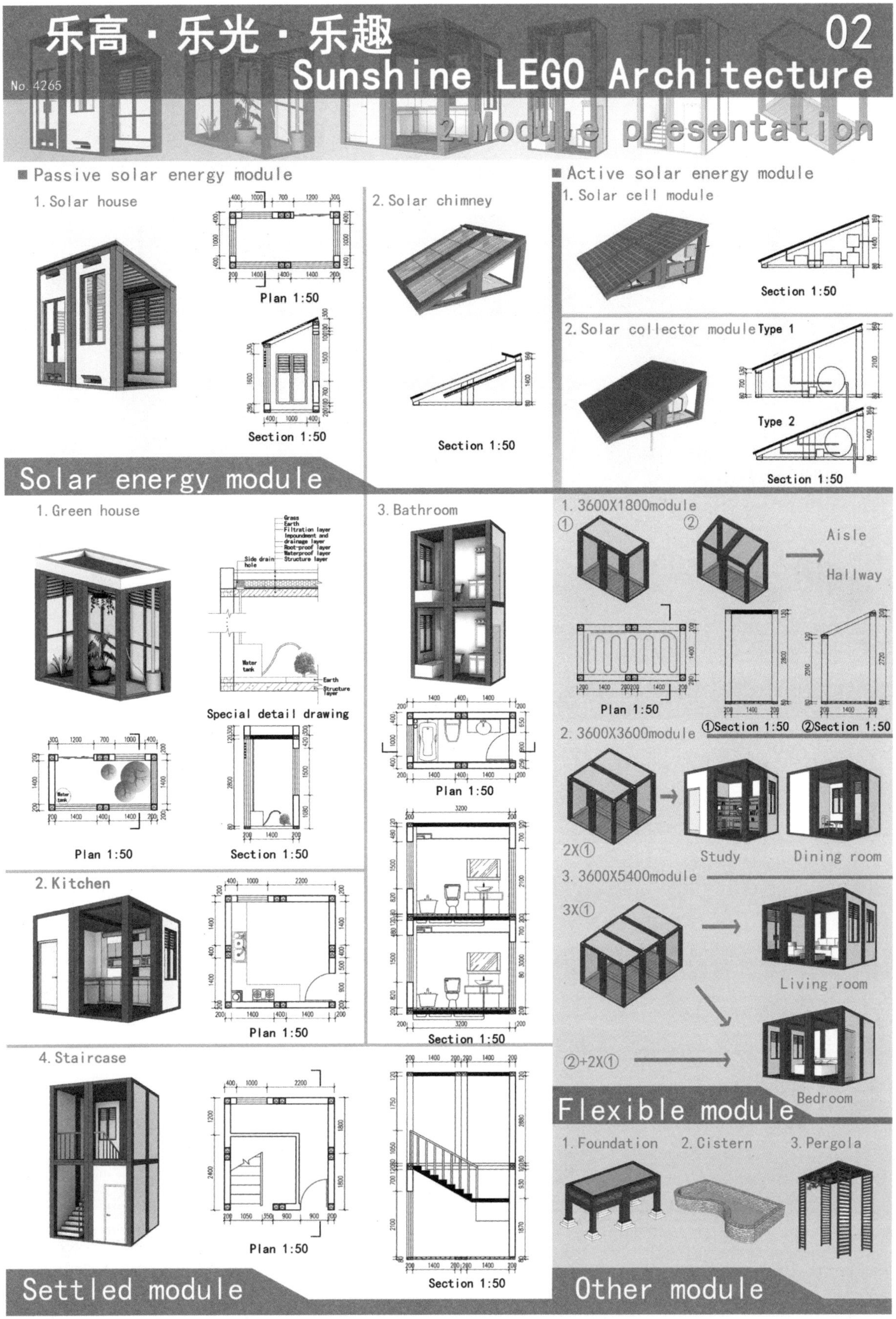

乐高·乐光·乐趣
Sunshine LEGO Architecture
2. Module presentation

■ Passive solar energy module

1. Solar house — Plan 1:50, Section 1:50
2. Solar chimney — Section 1:50

■ Active solar energy module

1. Solar cell module — Section 1:50
2. Solar collector module — Type 1, Type 2 — Section 1:50

Solar energy module

1. Green house — Special detail drawing (Grass, Earth, Filtration layer, Impoundment and drainage layer, Root-proof layer, Waterproof layer, Structure layer, Side drain hole, Water tank) — Plan 1:50, Section 1:50
2. Kitchen — Plan 1:50
3. Bathroom — Plan 1:50, Section 1:50
4. Staircase — Plan 1:50, Section 1:50

Settled module

Flexible module

1. 3600×1800 module — ① Aisle / Hallway, ② — Plan 1:50, ① Section 1:50, ② Section 1:50
2. 3600×3600 module — 2×① → Study, Dining room
3. 3600×5400 module — 3×① → Living room; ②+2×① → Bedroom

Other module

1. Foundation
2. Cistern
3. Pergola

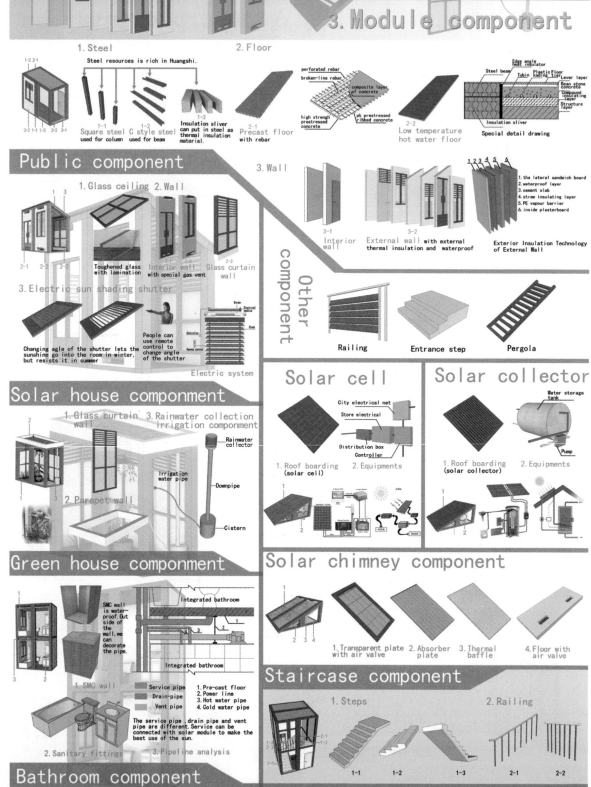

乐高·乐光·乐趣
Sunshine LEGO Architecture

04

4 Module combination

2015 台达杯国际太阳能建筑设计竞赛获奖作品集

No. 4265

1. Solar house + flexible module

Solar house can absorb heat from sunshine as more as possible and transfer heat to the flexible module. In summer, hot air in the flexible module also can discharge through the solar house.

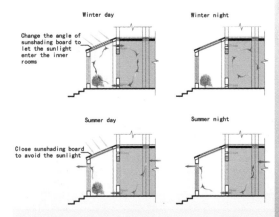

2. Green house or solar house + flexible module + foundation module with screen

On sunny days, the sunshine room absorbs a lot of solar radiant heat to heat the room. And redudant heat will be absorbed by screen thermal storage bed. At night, close the shutter and heat in screen thermal storage bed will release to keep the room warm.

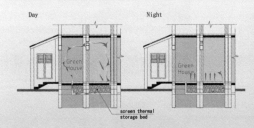

3. Solar chimney + flexible module

The sun radiation goes through the transparent glass plate into the chimney channel, then absorbed by heat storage material, heating the air inside the channel to produce the density difference between inside and outside. The process can complete the conversion from hot-pressing to wind pressure, makes the air inside the channel flow up.

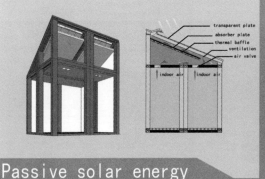

Passive solar energy

1. Solar cell module + flexible module

Roof module is connected with room module by the solar cell system. Solar cell electricity are used for equipment power consumption, incorporate into the common electrified netting, when solar energy battery are lack, we can use common electricity to need; if have richness can convey back to it.

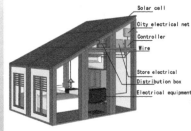

2. Solar collector module + flexible module

Roof module with the solar collector system can be connected with the hot water floor or toilet module by pipes. Water can be heated by solar collector, then, we can use hot water through hot water pipe.

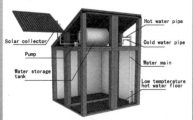

Active solar energy

Dampproofing

Foundation module + any other module

The damp-proof foundation can avoid dampness from the subsoil to keep the room dry. And at the bottom of foundation, the soil contains blast-furnace ash which is a by-product of steel can be used as anti-freeze soil padding.

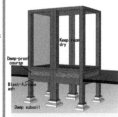

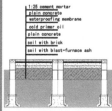

The damp-proof course of foundation

Utilization of water resources

1. Green house + other modules

On rainy days, rainwater run into the green roof of the green house module from slope roofs. After filtering of soil, rainwater run into the water storage tank through the water pipe. Water can be used to water trees and wash cars.

2. Cistern module + other modules
—— to form microclimate

The pool in the courtyard can collect rain and form the evaporation effect around the house to promote natural ventilation. It also can cool the wall outside the house and make the temperature lower in summer.

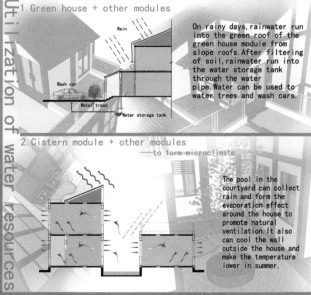

乐高·乐光·乐趣
Sunshine LEGO Architecture

No. 4265

05 — 5. Assembling process

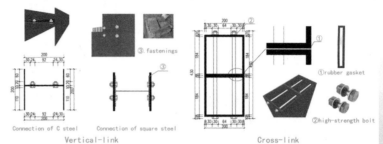

1. Connect steel structure

Connection of C steel — Vertical-link
Connection of square steel
③ fastenings
① rubber gasket
② high-strength bolt
Cross-link

2. Install the wall
① rubber gasket
② high-strength bolt
③ fastenings

Rubber gasket can be put between wall and steel to prevent water infiltration. Fastenings can be used to connected steel and wall.

Connections of steel and wall

3. Install the floor

The floor with rebar can be put into the prefabricated steel with holes, then we use the high-strength bolts to connect them.

Connections of steel and floor

4. Install special component of each module

Install stairs of staircase — The steps with rebar can be put into the prefabricated steel with holes, then we use the high-strength bolts to connect them. Steel welding can connect the steps firmly.

 Install solar cell board of solar cell module
 Install solar collector board of solar collector module
 Install intergrated bathroom of bathroom module
 Install water pipe of bathroom and kitchen
...

In the factory
Step1: make modules

Step2: transport modules in the factory to the construction site

Factory → Transportation → Construction site

Step3: foundation construction

Foundation trench

Step4: assemble modules

Step5: strengthen collection between modules

In the construction site

Construction waste — Less pollution
Building workers — Crane
Construction site — Factory
Slow — Fast
Traditional building — Fabricated building

Advantages of fabricated building

1. Prefabricated buildings can save resource and have nearly no pullution, which is good to the environment.
2. Components and modules can be lifted with crane, so less workers are needed, which can save the cost.
3. Because all the components and modules can be produced in the factory, the building process is controlled by the weather rarely.
4. The speed of prefabricated construction is fast, especially modular architecture. Maybe the process only needs one day.

Compare

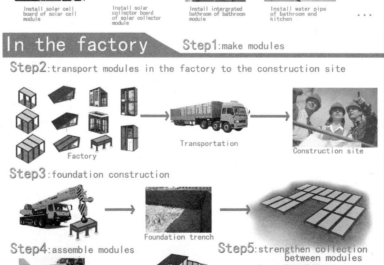

Flexible module: ¥650 ¥400 ¥1000 ¥1300
Solar cell module ¥4000
Solar collector module ¥5200
Foundation ¥400

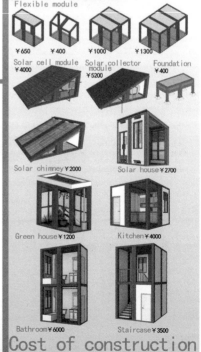

Solar chimney ¥2000
Solar house ¥2700
Green house ¥1200
Kitchen ¥4000
Bathroom ¥6000
Staircase ¥3500

Cost of construction

乐高·乐光·乐趣
Sunshine LEGO Architecture

No. 4265

06

2015台达杯国际太阳能建筑设计竞赛获奖作品集

6. Assembling results

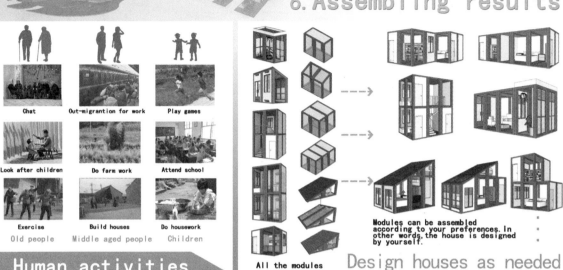

Human activities — Old people / Middle aged people / Children
(Chat, Out-migration for work, Play games, Look after children, Do farm work, Attend school, Exercise, Build houses, Do housework)

All the modules — Design houses as needed

Modules can be assembled according to your preferences. In other words, the house is designed by yourself.

1st floor plan 1:150   2nd floor plan 1:150

This assembling result is suitable for family of three people or the old couple. They can enjoy their own courtyard and have a quiet life.

Scheme one

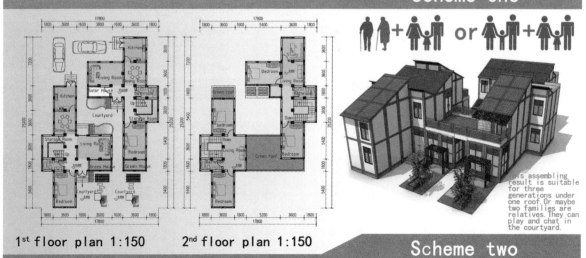

1st floor plan 1:150   2nd floor plan 1:150

This assembling result is suitable for three generations under one roof. Or maybe two families are relatives. They can play and chat in the courtyard.

Scheme two

综合奖·三等奖
General Prize Awarded · Third Prize

注　册　号：3553
项目名称：片山屋（青海）
　　　　　Sliced Rockery House
　　　　　(Qinghai)
作　　者：牛微、王楠、邵月婷、
　　　　　崔雅婧、孙楠、张玉洁、
　　　　　韩昆衡、赵若婉
参赛单位：山东建筑大学

专家点评：

作品规划布局紧凑，建筑平面功能分区合理，造型借鉴了当地传统的庄窠形式，并运用现代技术延伸了传统建筑的理念。采用聚光型太阳能集热板及储热系统。其是利用卵石床的白天集热、夜间散热的构造具有可操作性。利用太阳能生物质能联合沼气池的供热供气技术为厨房提供烹调热源，营造了舒适的生活环境。在建筑材料方面将当地材料转化为新型夯土墙结构。作品的建筑外形设计深度不足。

This is a compact design, with reasonable plane zoning; it draws on local culture and the use of modern technology to extend the concept of traditional architecture. This design uses a condensing type solar collector and thermal storage system; in particular, the gravel bed provides efficient day time heating and night time cooling. The use of solar and biogas, combined into a biogas tank, provides thermal technology that even provides heat for cooking; creating a very comfortable living environment. With regard to construction materials, local materials were transformed into new compact earth wall structures. However, the depth of the architectural design seems insufficient.

片山屋 SLICED ROCKERY HOUSE
INTERNATIONAL SOLAR BUILDING DESIGN COMPETITION 2015

Planning

▢ Simulation Analysis

Using CSWD meteorological data, Huangyuan county is located at 36.32 degrees north latitude, 101.08 degrees east longitude. In order to make full use of solar energy, this scheme building orientation is from South West 10 degrees to South East. Taking into account the river landscape sight, the choice is between the South East 0-10 degrees.

From the figure:
Best orientation: South West 5 degrees
Suitable orientation: SW-SE 30 degrees

Wind resistant performance: 0,0,0 degrees
Wind resistant performance: 0,5,10 degrees (Selection priority)

Wind resistant performance: 0,0,0 degrees
Wind resistant performance: 10,10,10 degrees

▢ Site Analysis

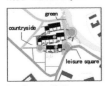 

Road — Classification system
View — A good view of the landscape
Landscape Node
Anti-wind — Stop the four winds

Architecture

▢ Concept Formation

The local architectural style
Local traditional residential houses are courtyard buildings based on earth

1. Traditional Zhuangke consists of functional rooms and courtyards. For example, scripture hall, living room, bedroom.
2. In the scheme, the ground floor retains the traditional living forms and compound relations to consistent with the local people's living habits.
3. On the basis of retaining the traditional part, the building integrates modern way of life.

4. Add residential space above the traditional part. The south is the bedroom, north is recreation area and bath room.
5. South solar room make the light energy into heat energy, dealing with the cold climate.
6. Ram-earth tube solve the problem of ventilation and lighting.

▢ Feature

Eave / Gable / Door / Window / Roof / Scripture hall Facility

Huangyuan county is the Han and Tibetan inhabited areas, buildings have Tibetan characteristics. Architectural details such as the roof, doors have local characteristics. Scripture hall is important in daily life.

▢ Ram-earth Language

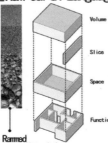

Volume / Slice / Space / Function

Rammed earth Material

▢ Traditional Construction Technology

Local traditional building materials mainly are: brick, ram-earth, ram-earth type material, grass mud and so on. Traditional construction technology is rich and unique.

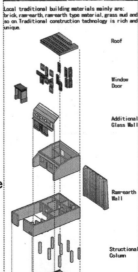 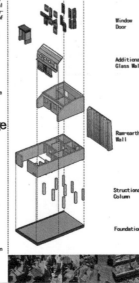

Roof / Window / Door / Additional Glass Wall / Ram-earth Wall / Structional Column / Foundation

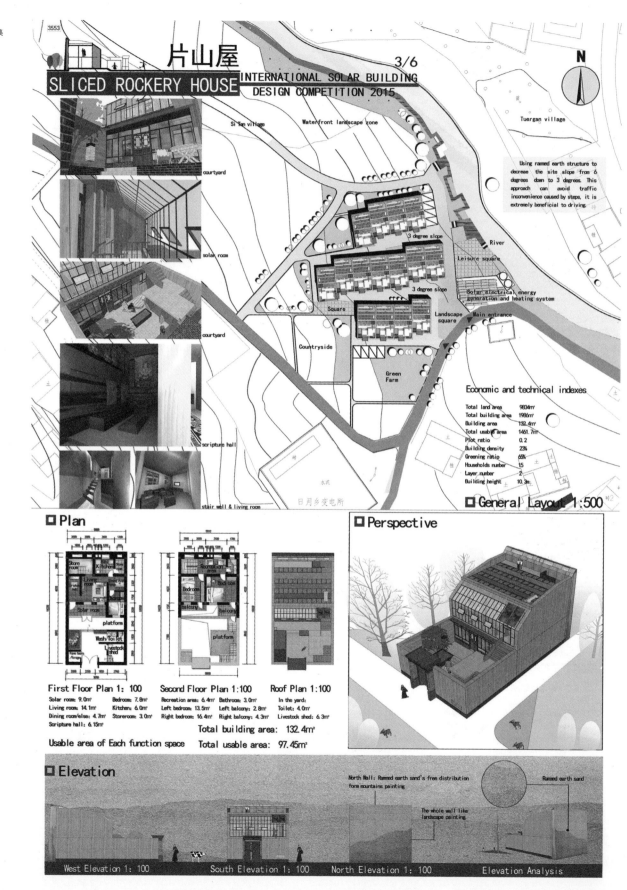

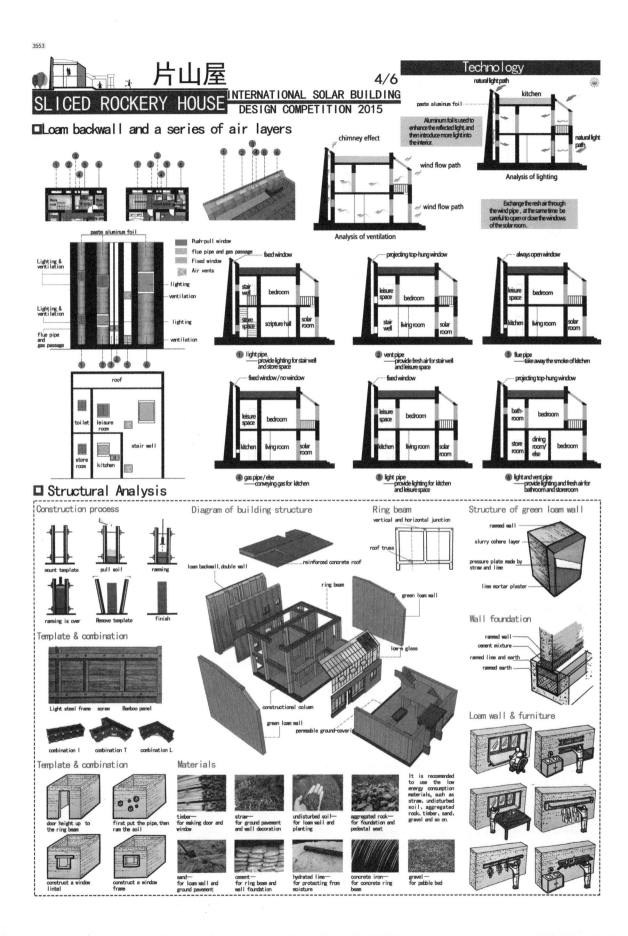

片山屋 / SLICED ROCKERY HOUSE

Energy-saving Technology

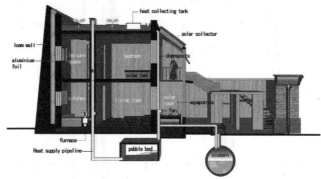

Solar Room、Pebble Bed and Furnace

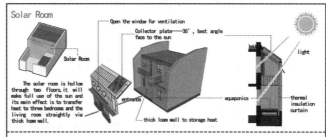

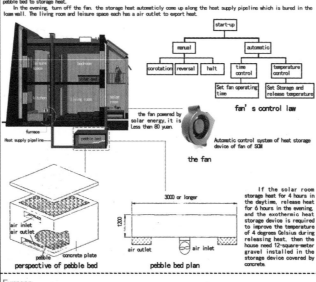

Furnace

At the same time, there is a furnace connected to the heat supply pipeline, which is suggested to use in the winter. Once fire up the furnace, the hot air will release heat to the living room downstair and the leisure space upstairs, as well as storage heat in the pebble bed for the night.

USE pattern 1 — In the pipeline obligate air outlets towards the living room and the leisure space.

USE pattern 2 — Connect the pipeline with radiator in the living room and the leisure space.

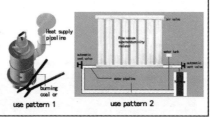

use pattern 1   use pattern 2

Roof Greening

Because the local has a low rainfall, so the local buildings always have earth covered roofs to keep the rooms warm.
In the structure of the planted roof, there is a drainage layer to storage water and deliver water to the cistern through the downspouts.

Rain Water Collection

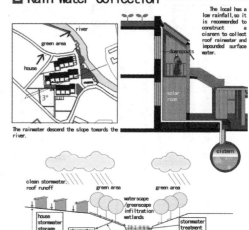

Solar Water-heating System

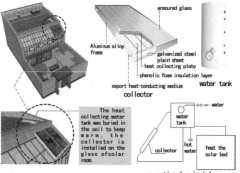

Solar bed

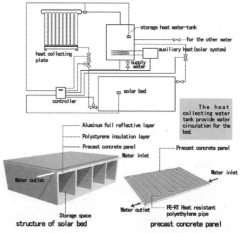

SLICED ROCKERY HOUSE
片山屋
INTERNATIONAL SOLAR BUILDING DESIGN COMPETITION 2015

◘ Solar Electrical Energy Generation and Heating

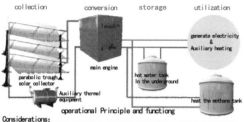

operational Principle and functions

Considerations:
- close to the methane tanks, keep them warm and running normally.
- Set in the place where is full of sunshine and no shelter.
- Place the tracking collectors towards 5 degrees south by east.
- In addition to heat the methane, the heat are also used for heating the beds in each house.

Technical Parameter

Heating area	1400 m²
Heat medium	320#Heat conduction oil
The Heat collector tracking mode	The North-South track
The collector daylighting area	150 m²
The collector cover area	300 m²
The driving temperature	50~20°C
Thermal storage material	H₂O
The temperature of beds	35~40°C
The temperature of methane tanks	37~53°C

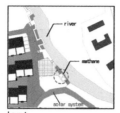

Layout

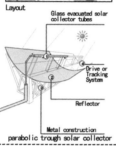

parabolic trough solar collector

◘ Door & Window

Design discipline:
Frame material: timber
coefficient of heat transfer K≤1.3W/(m²·K)
Glass: coated glass, Low-E glass
Three-layer sealing materialr

Reducing partition of door and windows can be convenient for manufacture and can also effectively enhance the thermal insulation performance.

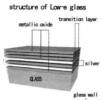

structure of Low-e glass

wooden door — wooden window — windows of solar room

Function of Low-e glass

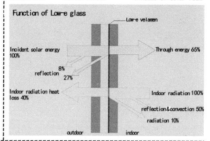

◘ Solar Methane

- Methane: the fermenting mass generates gas, which fuels stoves and replaces kerosene.
- It is recommended to build a cattle and sheep farms if the village want to use methane as main cooking fuel.
- Built the methane tank in the lower reaches of the river and the low place of the site.
- 18-cubic-metre methane tank is enough to product methane for the 15 families.

Solar collector and boiler are used to heat the fermented liquid in winter.

Operational Principle

Advantages:
- Improve agricultural ecological environment
- increase fertilizer for the farmland
- promote the yield of crops, and product of green food.
- save labor and capital
- significantly improve soil structure
- promote the development of livestock and poultry breeding

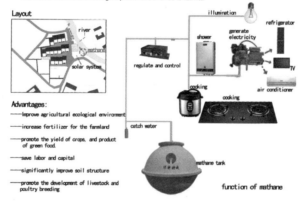

function of methane

◘ Shower Water Collection

Take a shower water collection, and use the collected water to washing the toilet.

Advantages:
water conservation
Low cost and fast-speed

◘ Aquaponics

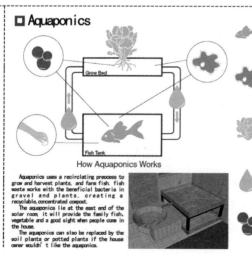

How Aquaponics Works

Aquaponics uses a recirculating precoess to grow and harvest plants, and farm fish. fish waste works with the beneficial bacteria in gravel and plants, creating a recyclable, concentrated compost.
The aquaponics lie at the east end of the solar room, it will provide the family fish, vegetable and a good sight when people come in the house.
The aquaponics can also be replaced by the soil plants or potted plants if the house owner wouldn't like the aquaponics.

Fish are fed food and produce Ammonia rich waste. Too much waste substance is toxic for the fish, but they can withstand high levels of Nitrates.

The bacteria, which is cultured in the grow beds as well as the fish tank, breaks down this Ammonia into Nitrites and then Nitrates.

Plants take in the converted Nitrates as nutrients. The nutrients are a fertilizer, feeding the plants. Also, the plant roots help filter the water for the fish.

Water in the system is filtered through the grow medium in the grow beds. The water also contains all the nutrients for the fish.

Oxygen enters the system through an air pump and during dry periods. This oxygen is essential for plant growth and fish survival.

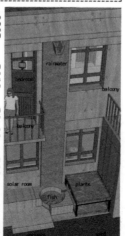

综合奖・三等奖
General Prize Awarded · Third Prize

注 册 号：3660

项目名称：日月生辉——青海日月乡低能耗住房（青海）
The Brightness of Riyue—
Low-energy Housing Project in Riyue of Qinghai (Qinghai)

作　　者：许泽寰、王雪菲、薛小刚、林蓓蓓

参赛单位：西安建筑科技大学、上海交通大学

专家点评：

作品规划富有层次，村落的空间营造富有特点，划分出不同尺度和功能的公共空间，并对用地内的建筑布局进行了细致的推敲。建筑平面布局符合农牧民生产和生活需求，并针对不同的生活方式设计了多种户型。建筑外形符合藏区住宅的建筑特色，村落内外公私空间有趣而又富于变化，有助于形成村落共同意识。作品在太阳能主、被动技术运用方面表达不足。

This design is eloquently thought out and planned; the village space is rich with features, and public space use is divided into various different scales and functions. The building layouts are in line with the needs of local farmers; and a variety of units are available for production and living in many different ways. Architectural appearance conforms to Tibetan residential features; with interesting public and private spaces both inside and outside the village; this gives a sense of common awareness to its residents. However, this design is lacking in the use of passive and active solar technology.

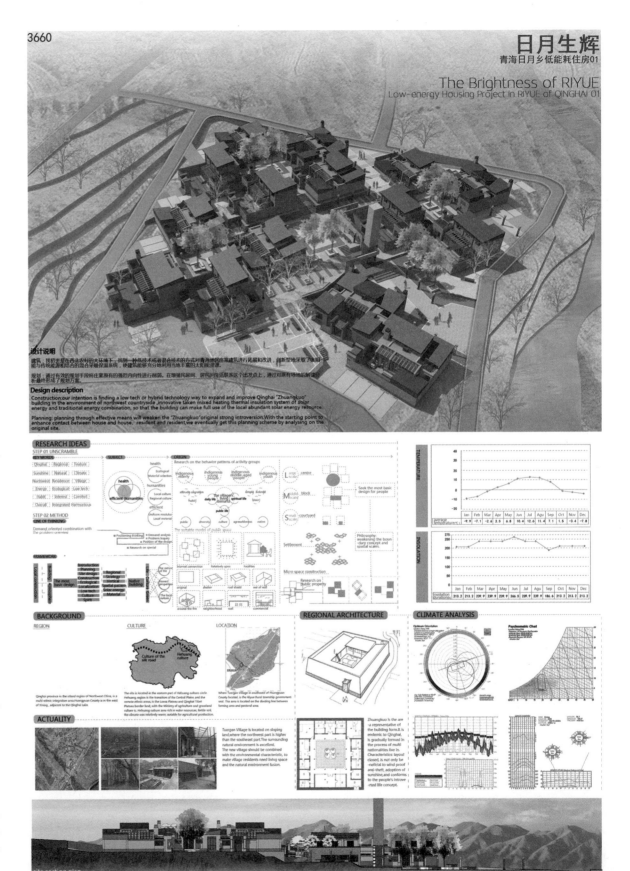

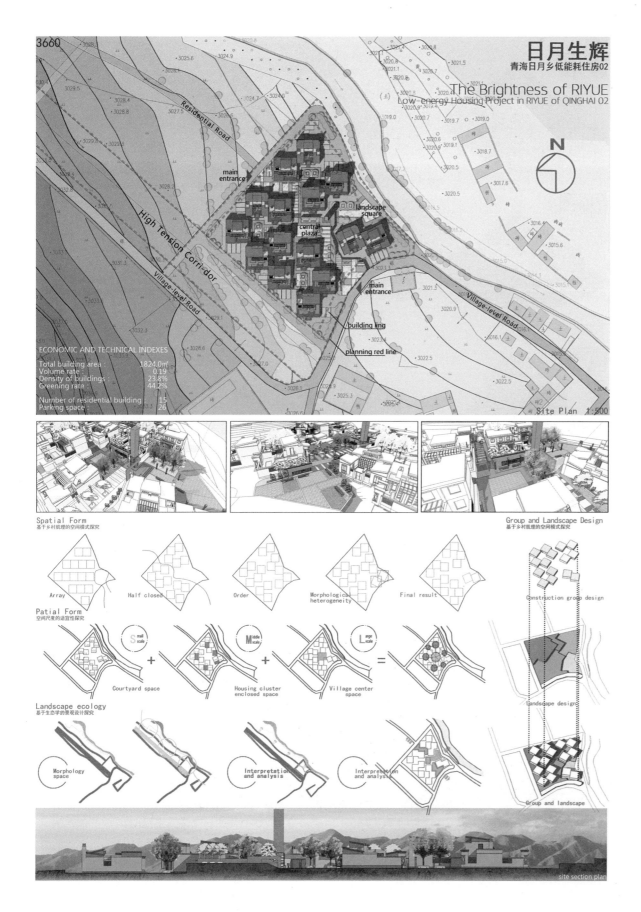

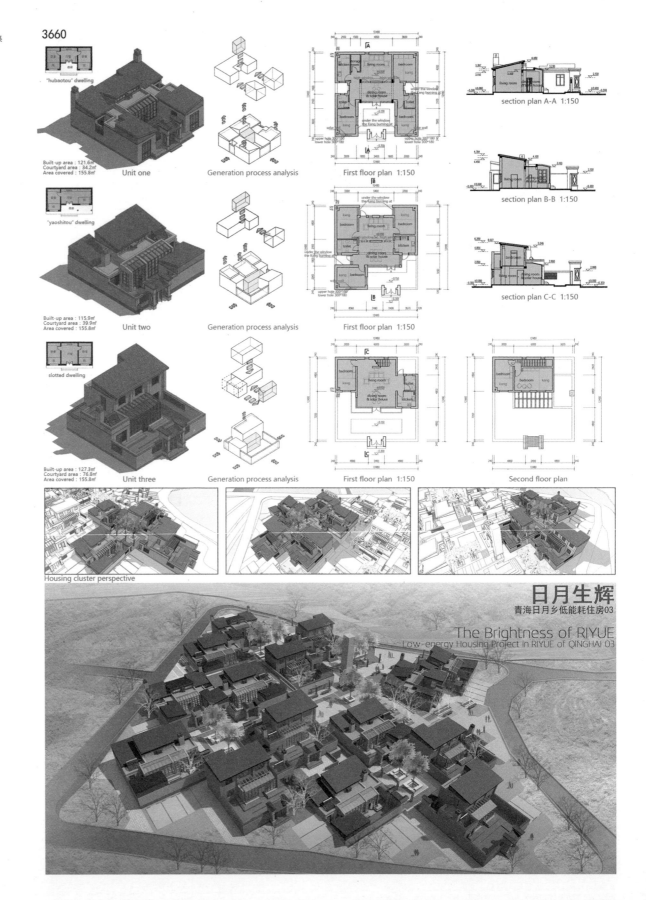

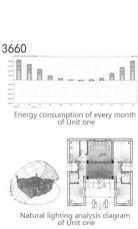
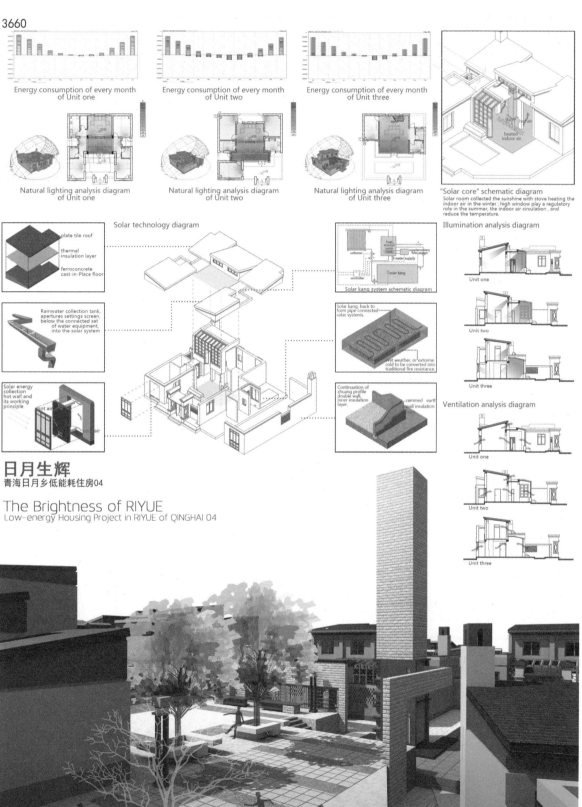

日月生辉
青海日月乡低能耗住房04

The Brightness of RIYUE
Low-energy Housing Project in RIYUE of QINGHAI 04

综合奖·三等奖
General Prize Awarded · Third Prize

注 册 号：3677
项目名称：庄窠·融光（青海）
　　　　　Zero-carbon · Adaptable Sustainable Residential Design (Qinghai)
作　　者：高力强、加晶晶、仇朝兵、
　　　　　兰　亮、庄冠存、刘　其、
　　　　　李　健、Jian Zuo
参赛单位：石家庄铁道大学、天津大学、
　　　　　University of South Australia

专家点评：

作品运用了当地传统建筑设计手法，应用当地建筑材料和构造方式。采用了太阳能发电、太阳能热水、太阳能沼气、太阳能炕等主动式太阳能系统，以及雨水利用、热回收功能通风塔、绿化遮阳等节能技术，并将两者与建筑进行了一体化设计，太阳能、低碳技术与藏族特色的建筑造型结合巧妙。作品在规划方面设计深度不足。

This design uses local traditional architectural design practices, as well as application of local building materials and construction methods. It uses many active solar energy system, including solar power generator, solar water heater, solar energy and biogas, solar Kang, etc.; it also incorporates rainwater utilization, heat recovery ventilation towers, green shading and other energy saving technologies. By integrating these 2 systems, they've achieved construction of an optimal design. Lastly, solar energy and low carbon technology are well in tuned with local Tibetan architectural characteristics. However, the depth of the architectural design seems insufficient.

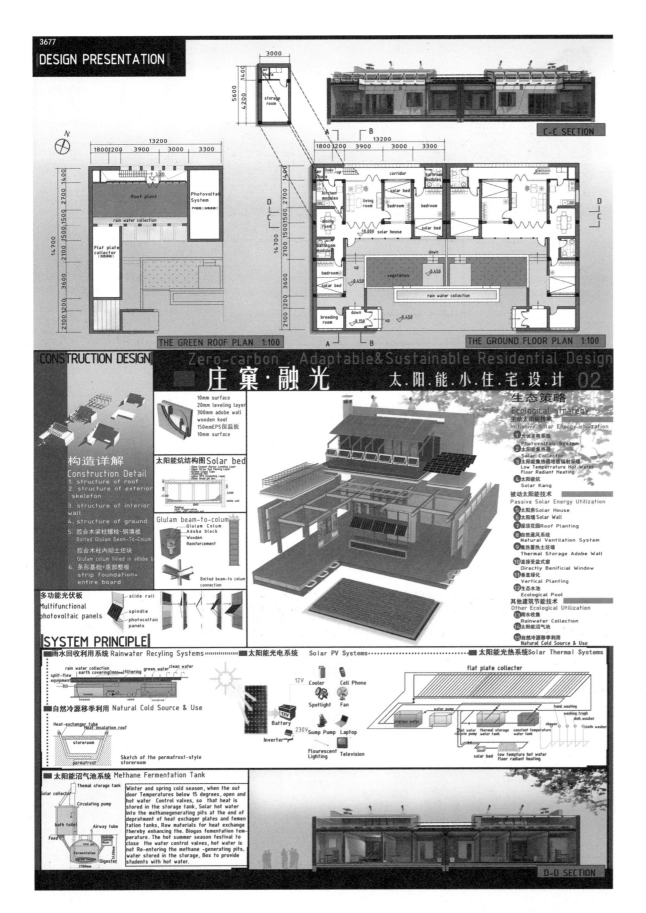

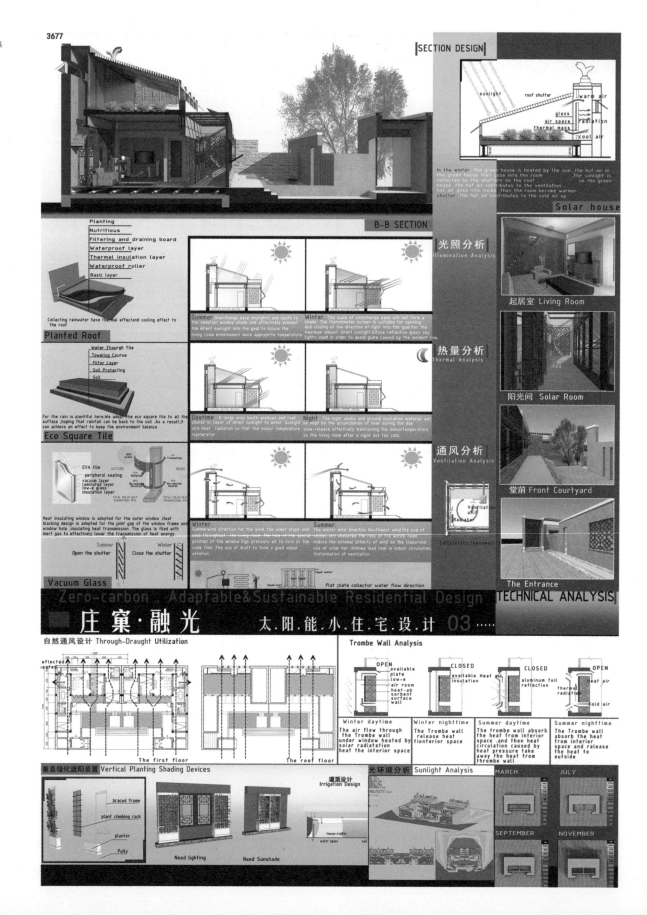

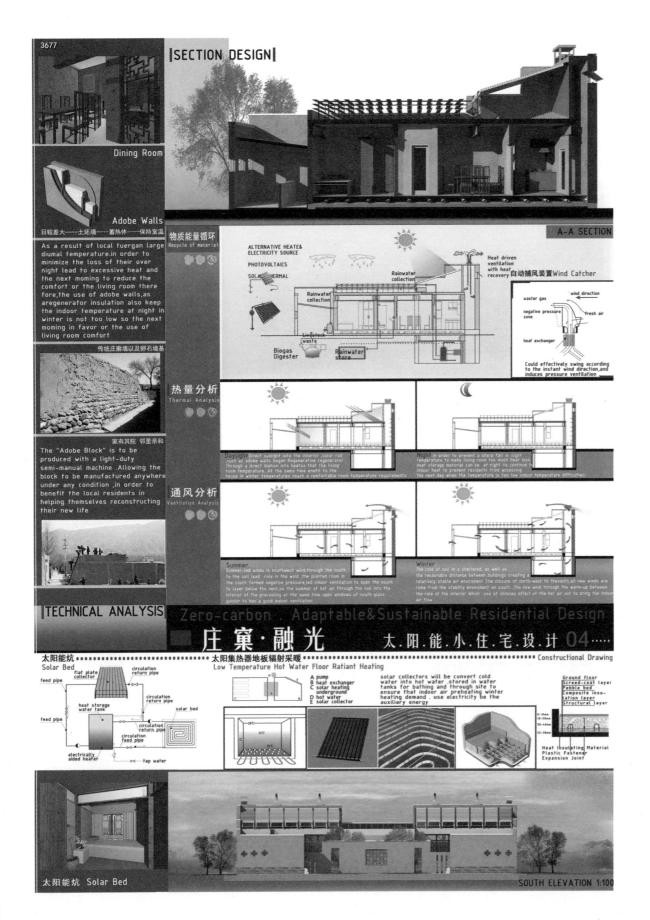

综合奖·三等奖
General Prize Awarded · Third Prize

注 册 号：3704
项目名称：山间"屯"光（青海）
　　　　　Indoor Sunshine and Breeze (Qinghai)
作　　者：高力强、孟凡平、魏智强、
　　　　　张晶石、季思雨、尹 欣、
　　　　　高 洁、朱相栋
参赛单位：石家庄铁道大学、河北工业大
　　　　　学、清华大学

专家点评：

作品将绿地集中规划布置，场地与地形结合较好。建筑平面功能布局合理、空间丰富，建筑外形将现代技术与传统建筑元素相结合，立面造型有利于太阳能的吸收。被动与主动太阳能利用结合合理，同时考虑了冬季保温、夏季自然通风、光热与光电综合利用等，具有较强的可实施性。作品规划布局的灵活性不够。

This design focuses on the layout of green space and maintains an excellent balance between the ground and terrain. The layout foundations are spaced rationally, and features modern architectural technology; this combine with the surface design makes it very conducive to absorbing solar energy. Passive and active solar energy combination and utilization is reasonable, taking into account insulation during winter and natural ventilation during summer; in addition, there is comprehensive utilization of solar thermal and photovoltaic generation. This design can be implemented easily, but the layout's flexibility is somewhat lacking.

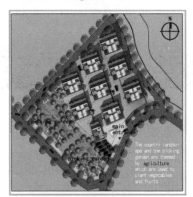
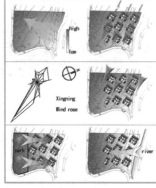
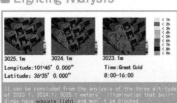

First Floor Plan 1:100 ## Second Floor Plan 1:100

1. Storage room (2.6㎡)
2. Restaurant (10.0㎡)
3. Kitchen (7.2㎡)
4. Toilet
5. Secondary room (11.0㎡)
6. Saloon (21.4㎡)
7. Entrance (5.0㎡)
8. Zen style (4.4㎡)
9. Plant room (16.5㎡)
10. Master room (13.5㎡)
11. Secondary room (12.0㎡)
12. Sunshine Room (13.5㎡)
13. The balcony
14. Pool
15. Header tank
16. Feed inlet
17. Feed outlet
18. Cellar (10.5㎡)
19. courtyard

Perspective

Countyard
Balcony

Cellar

Lighting Analysis

spring equinox | summer solstice | autumnal equinox | winter solstice

Economic Index

- Total area: 117.6㎡/house
- Covered area: 151㎡
- Planning building number: 17
- Site area: 960㎡
- Plot area: 5000㎡
- Picking garden area: 2500㎡
- Volume fraction: 40%
- Green area: 4500㎡
- Greening rate: 45.8%

Perspective

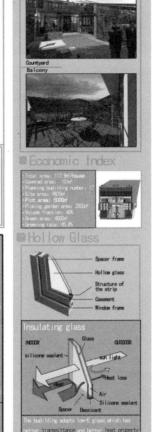

First floor

Second floor

The wall between two families is sealed after it is filled with rock wool

Hollow Glass

- Spacer frame
- Hollow glass
- Structure of the strip
- Casement
- Window frame

Insulating glass

INDOOR / OUTDOOR
silicone sealant, Glass, sun light, Heat loss, Air, Silicone sealant, Spacer, Dessicant

The building adopts low-E glass, which has better transmittance and better heat property

山间 "屯" 光   INDOOR SUNSHINE AND BREEZE
Solar Energy Residential Design 02

Photovoltaic System

Solar Cells, Junction Box, Male Connector, Female Connector, Cable Assembly, Batteries, Controller, Inverter, Branch Connector, DC load, AC load

Solar Panel Structure

Toughened Glass / EVA / CITL / EVA / TPT/BBF

Computational Process:
- Solar battery power: 150w/㎡·day
- Receive radiation: 4kwh/㎡·day
- Conversion ratio: about 15%
- Producing power: 4*15%=0.6kwh/day ≈0.6㎡/day
- Power consumption: about 3.6㎡/day
- Area: 3.6/0.6=6㎡

Electricity balance
- Producing power
- Power consumption

Floor and Wall

1. The vertical connection
2. keel
3. plasterboard
4. flax
5. rock wool
6. precut loam wall
7. wood floor
8. coating
9. keel
10. precast floor slab
11. vent
12. suspended ceiling
13. air duct
14. floe

Precut Loam Wall

Flax / Flax Straw and clay / The connecting hole

Local materials:
1. Flax 2. Clay 3. Wood Scraps

Advantage:
1. Get material convenience
2. The rapid assembly
3. Good heat preservation performance

A-A Profile 1:100

Cellar / Methane tank

Behavioural Analysis

Sleep — Get up — Breakfast — Work — Lunch — Work — Dinner — Watch TV — Sleep

— Behavior strength

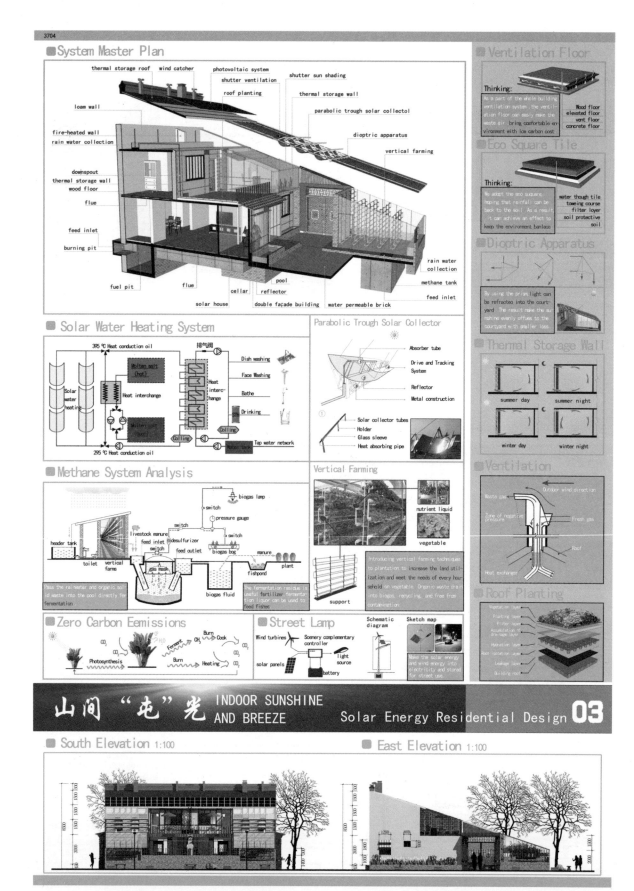

山间"屯"光 INDOOR SUNSHINE AND BREEZE — Solar Energy Residential Design 04

Planting
1. Sunshade
2. Rimp
3. Isolate noise
4. Photosynthesis

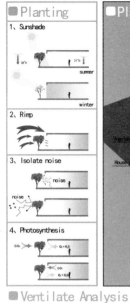

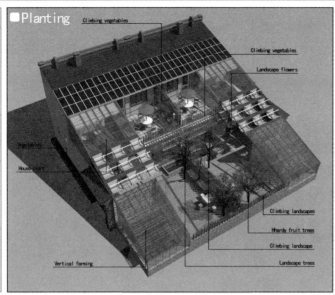

Indoor

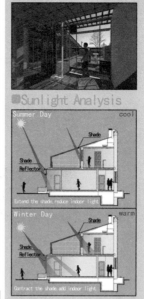

Sunlight Analysis

Summer Day — cool: Extend the shade, reduce indoor light.

Winter Day — warm: Contract the shade, add indoor light.

Shading Analysis

open / pack up / close

- Open the wooden louvres, close the curtain, the sunlight come into indoor. WARM
- Close the wooden louvre, open the curtain and breeze across the salon house. COOL
- On the night, close the wooden louvre, open the curtain, has heat insulation, dust-proof effect. WARM

Ventilate Analysis

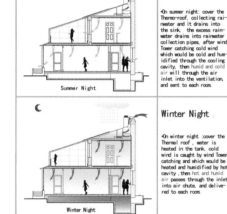

Summer Night: On summer night: cover the Thermo-roof, collecting rainwater and it drains into the sink, the excess rainwater drains into rainwater collection pipes, after wind Tower catching cold wind which would be cold and humidified through the cooling cavity, then humid and cold air will through the air inlet into the ventilation, and sent to each room.

Summer Day: On summer daytime: cover the Thermo-roof, collecting rainwater and it drains into the sink, the excess rainwater drains into rainwater collection pipes, after wind Tower catching hot wind which would be cold and humidified through the cooling cavity, then humid and cold air will through the air inlet into the ventilation, and sent to each room.

Winter Night: On winter night: cover the Thermal roof, water is heated in the tank, cold wind is caught by wind Tower catching and which would be heated and humidified by hot cavity, then hot and humid air passes through the inlet into air chute, and delivered to each room.

Winter Day: On winter daytime: open the Thermal roof, gas is heated in the heat Chamber and water is heated in the tank, after wind Tower catching cold wind which would be heated and humidified by hot cavity, then hot and humid air passes through the inlet into air chute, and delivered to each room.

Heating System

Thinking — How to use it? What can we do? HEATING FOR BUILDING!

Raw material: til straw, wheat straw, rape straw

Instructions: Before the winter comes, pulverize the stalk of crops into sawdust and put it in the inlet of 30m² fuel pits, pilot it but without producing fire, and by the means of controlling the air volume through air inlet to control burning speeds, to provide heat for buildings, and keep the temperature at 20 degrees Celsius approximately throughout the winter, which could save one 3-4t/winter.

Fuel pit plane analysis — feed inlet, air vents, fuel pit, flue, cellar

Model — chimney, flue, fuel pit, cellar

Advantages:
1. Low heating costs
2. Convenient materials management
3. Well-distributed heat
4. Clean and neat, safe and healthy

Solving problems:
1. Low temperature in winter
2. Straw without storages place
3. Environment contamination by burning straw
4. Wasting of resources

plane analysis — feed inlet, air vents, fire-heated wall, loam wall, solar house, thermal storage wall

Heating analysis / profile analysis — feed inlet, thermal storage roof, rain water collection, flue, fire-heated wall, fire resisting floor, fuel pit

Water System

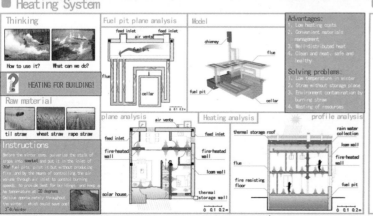

rain, river, water treatment, storage water tank, parabolic trough solar collector, plant, toilet, kitchen, eco square tile, feed inlet, rain water collection

Legend: nodal point, city water, hot water, rainwater collection, sewage

综合奖·三等奖
General Prize Awarded · Third Prize

注　册　号：3837
项目名称：八分宅（黄石）
　　　　　80% House (Huangshi)
作　　者：林海锐、刘小康、胡浩森、
　　　　　何傲天、刘　晶、周楚晗、
　　　　　朱嘉懿、许安江、周一冲
参赛单位：华南理工大学建筑学院青年设
　　　　　计工作室

专家点评：

作品沿用当地传统民居的布局形式，平面功能采用中轴对称的形制，功能分区清晰明确，满足使用需求。建筑立面造型简单且有利于太阳能被动技术应用，利用冬夏季太阳能高度角的变化实现了夏季遮阳和冬季采暖。作品充分考虑了太阳能光热和光伏的综合利用并与太阳能被动技术进行良好的结合，但在太阳能综合利用的设计方面略显不足。

This design uses a traditional layout based on local residents, and is axial symmetrical-shaped; function zoning is very clear and meets the requirements. The facade style is simple and beneficial to the application of passive solar energy; it utilizes the large changes in solar altitude between winter and summer to adequately insulate and ventilate with solar energy. Although this design takes full account of comprehensive solar thermal and photovoltaic utilization, but the design's overall solar energy use is slightly lacking.

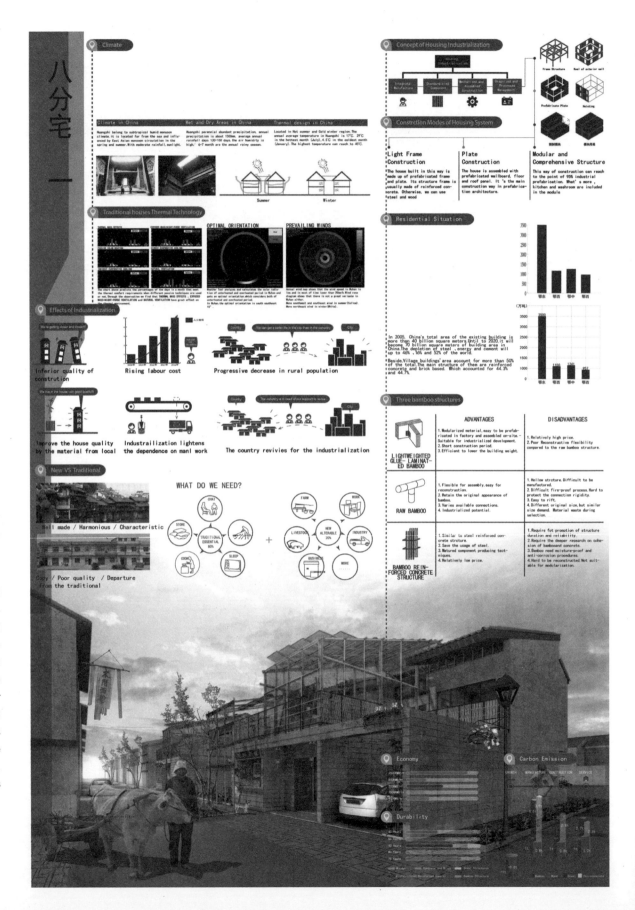

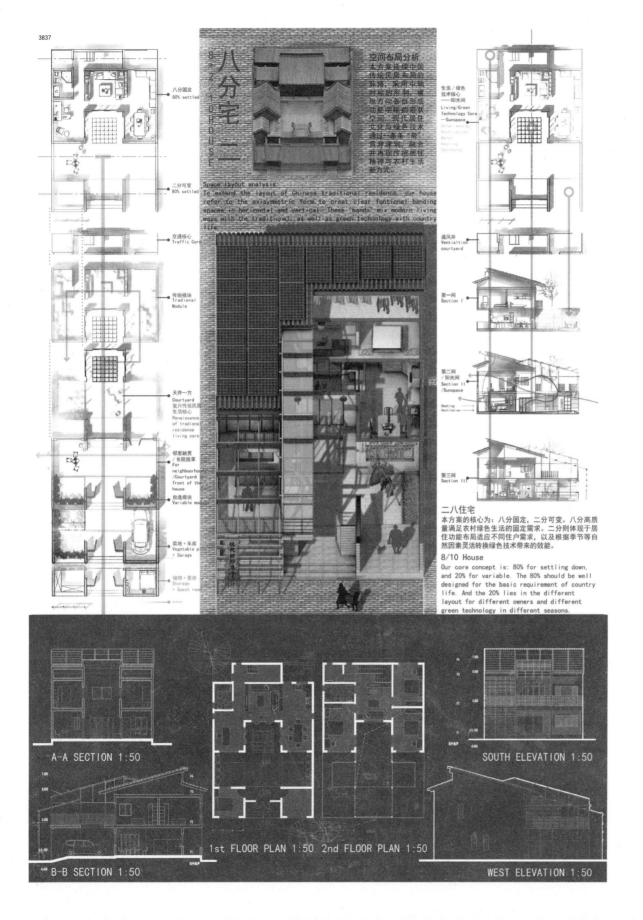

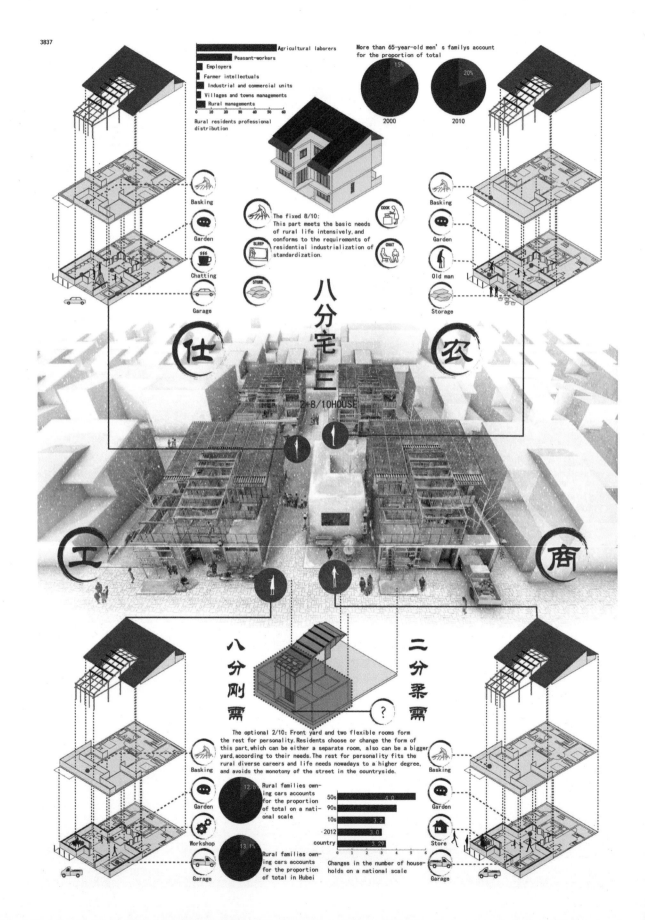

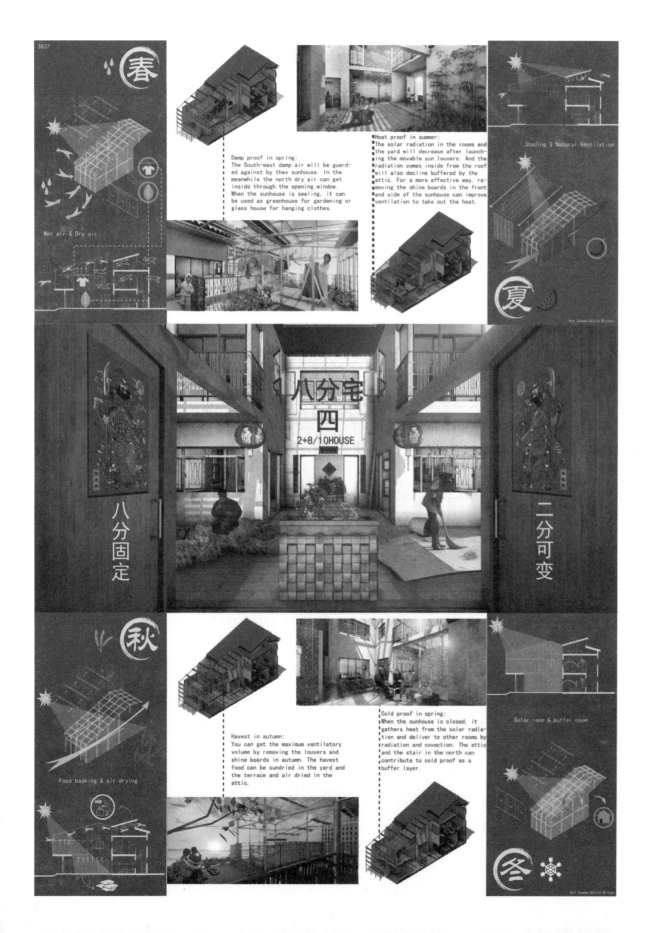

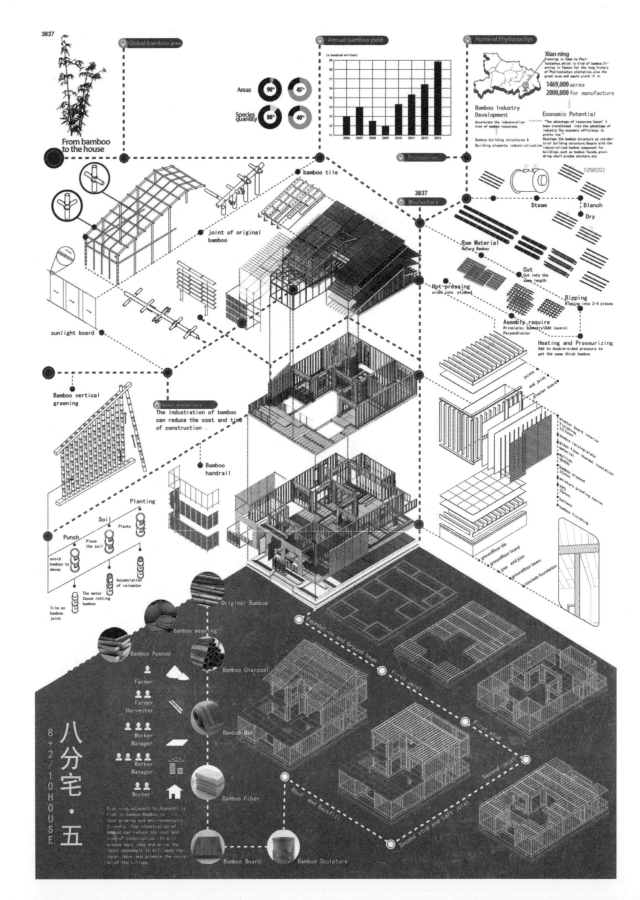

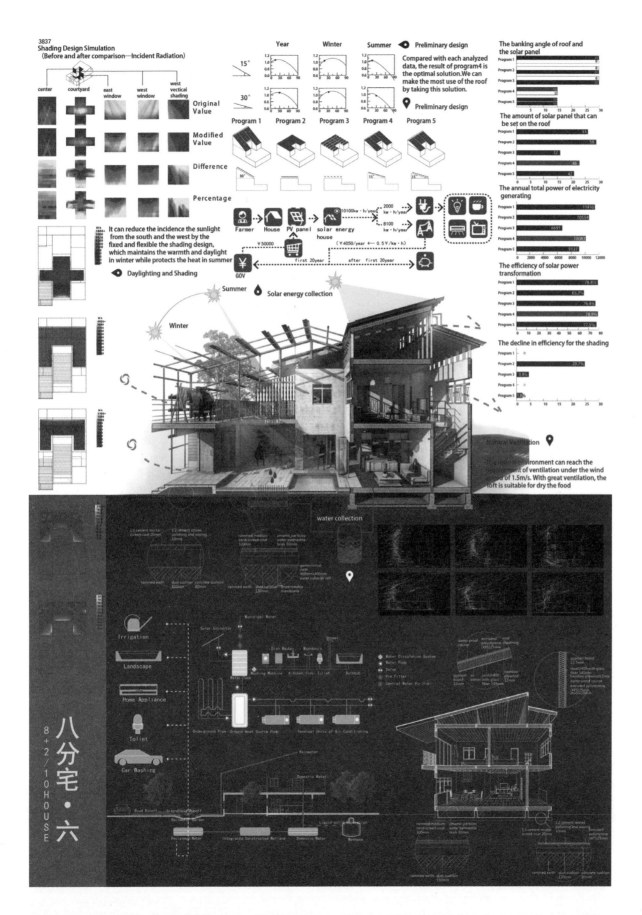

综合奖·三等奖
General Prize Awarded · Third Prize

注 册 号：4355
项目名称：墙·记忆（黄石）
　　　　　Wall Memory (Huangshi)
作　　者：王　欣、卫泽华、樊梦蝶、
　　　　　荆济鹏、张奕聪
参赛单位：北京交通大学

专家点评：

作品延续当地传统民居的建筑形式，提取封火山墙等传统建筑元素并与夏热冬冷地区的气候特点有机结合。建筑平面布局合理，室内空间利用紧凑。技术方面，在对室内热环境进行详细分析的基础上运用了被动太阳能集热、通风技术，部分解决了建筑冬季采暖和夏季隔热除湿问题；太阳能热水系统、雨水回收系统等主动技术进一步减少建筑运行能耗。作品采用预制装配式框架结构，施工速度快，可实施性强。作品的建筑立面设计深度不足。

This design continues the local traditions of architectural design; it uses traditional elements like fire gables to complement the local climate, which is hot in the summer and cold in the winter. The building layout is rational, and the interior space is utilized in a compact fashion. Technologically, based on a detailed analysis of a hot indoor environment, it utilizes passive solar energy and ventilation technology to partially solve the issue of heating in the winter and dehumidifying heat insulation in the summer. In addition, it uses solar hot water systems and a rainwater recovery system, along with other active technologies, to further reduce energy consumption. Lastly, the design uses preordered frames for construction, allowing the work to be done fast. However, this design's building facade lacks depth.

CHAPTER 2: APPLICATION

墙·记忆 *wall memory*

PLANE APPLICATION

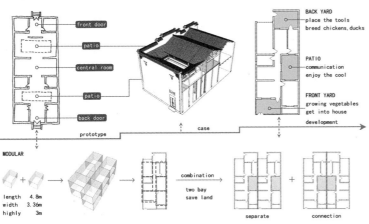

DESIGN SPECIFICATION

The design of rural housing in Huangshi area is founded upon the analysis of three factors: traditional architecture, climate and industrialization. We draw the conception of "WALL". "Fire gable" and courtyard are kept to cohere well with the tradition. Solar energy utilization and natural ventilation are taken seriously to meet the climatic adaptation. Fabricated and board formed lightweight steel structure is adopted due to its good economic performances, easy material availability and dry construction method.

黄石住宅公园农村住房设计，从延续传统民居特色以及夏热冬冷地区住房的气候适应性出发，结合产业化住宅的设计、建造问题，提炼出"墙"的概念。在建筑的外形和功能设计中，保留传统建筑中"封火山墙"、"天井"等语言；在气候适应性设计中，注重太阳能的利用以及自然通风的组织，解决南方地区夏季过热、冬季阴冷的问题；在住房产业化设计中，采用预制装配板式轻框架结构，其低造价、就地取材以及全干作业的特性均增加了项目的可操作性。

ECONIMIC INDICATIONS

TOTAL COVERY AREA	156.03㎡
TOTAL CONSTRUCTION AREA	249.81㎡
VOLUME RATIO	1.6
FIRST FIOOR CONSTRUCTION AREA	125.57㎡
BUILDING HEIGHT	9.11m

TECH APPLICATION

SOLAR POWER | VENTILATION | RAINWATER | BIOGAS | GEOTHERMAL

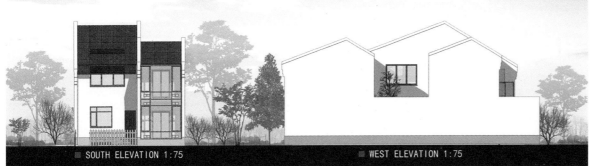

SOUTH ELEVATION 1:75　　　WEST ELEVATION 1:75

CHAPTER 4: TACTICS

PASSIVE VENTILATION AND HEATING IN WINTER

PASSIVE VENTILATION IN SUMMER

墙·记忆 *wall memory*

DRAINAGE GAS ACTICS ---------- DAYLIGHTING VENTILATION

DAYLIGHTING ANALYSIS

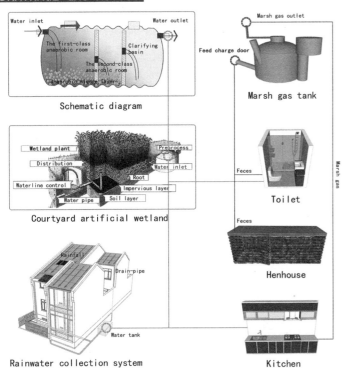

Summer Solstice 9:00 (morning) | Summer Solstice 15:00 (afternoon) | Winter Solstice 9:00 (morning) | Winter Solstice 15:00 (afternoon)

B-B section Daylighting
Indoor daylighting is good

C-C section Daylighting
In addition to the garage, daylighting is good

Daylighting First floor

VENTILATION ANALYSIS

Summer First floor | Summer Second floor | Winter First floor | Winter Second floor

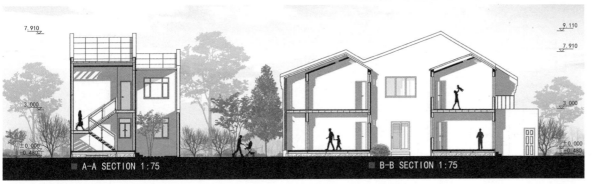

A-A SECTION 1:75   B-B SECTION 1:75

CIRCULATORY SYSTEM

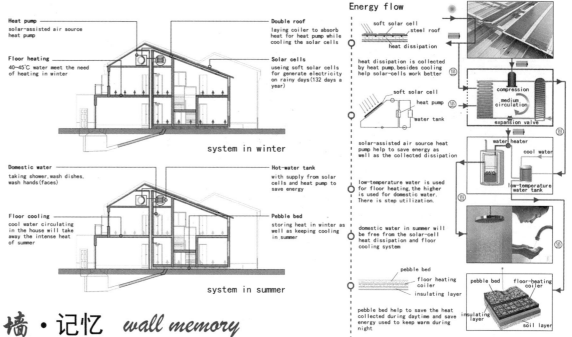

CHAPTER 5: TECHNIQUES

墙·记忆 *wall memory*

TECHNOLOGICAL ANALYSIS

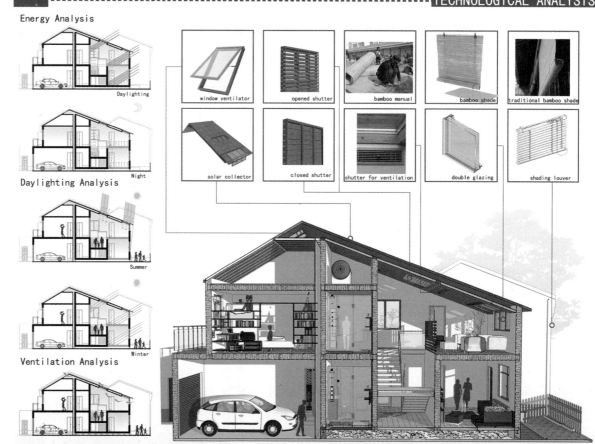

CHAPTER 6: INDUSTRIALIZATION

240 MODULUS

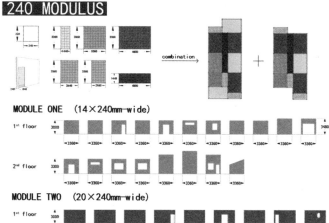

MODULE ONE (14×240mm-wide)
1st floor
2nd floor

MODULE TWO (20×240mm-wide)
1st floor
2nd floor

MODULE THREE (6×240mm-wide)

墙·记忆 wall memory

CONSTRUCTION

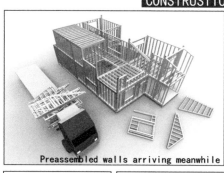
Preassembled walls arriving meanwhile

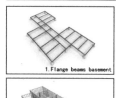
1. Flange beams basement

2. Precast concrete floor

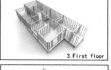
3. First floor

4. Floor slab

5. Second floor

6. Roof frame

STRUCTURE — FOUNDATION

Light-frame (cold-formed steel) walls sheathed with wood structural panels, rated for shear resistance or steel sheets.

150 / 62   146 / 58   200 / 75   196 / 71   100 / 50   96 / 46

DETAILS

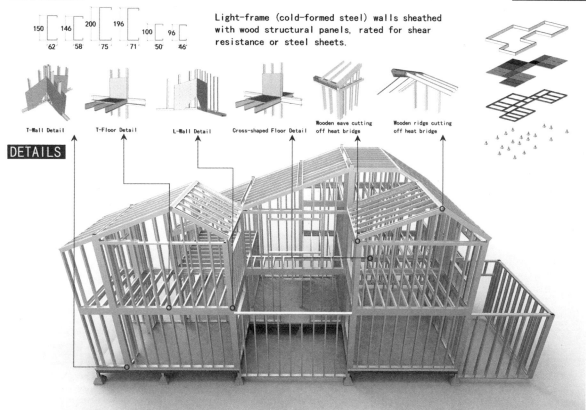

T-Wall Detail | T-Floor Detail | L-Wall Detail | Cross-shaped Floor Detail | Wooden eave cutting off heat bridge | Wooden ridge cutting off heat bridge

综合奖 · 优秀奖
General Prize Awarded · Honorable Mention Prize

注 册 号：3365
项目名称：生态藏居（青海）
　　　　　Ecological Tibetan Dwelling
　　　　　(Qinghai)
作　　者：诸葛文斌、郭梦露
参赛单位：西安科技大学

生态藏居 Ecological Tibetan Dwelling
农牧民定居青海低能耗住房项目设计　1

QINGHAI　　HUANGYUAN　　SITE

Traditional Reference

Architectual Forms　"口" type Courtyard Plane　"Zhuangke" Residential Plan　Altar

Entrance Door Style　Courtyard Greening　Eaves & Woodcarving　Plane Roof (for drying crops, etc.)

Passive Design Techniques' Influence on Comfort Percentages

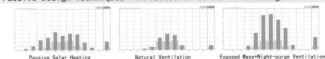

Passive Solar Heating　　Natural Ventilation　　Exposed Mass+Night-purge Ventilation

The light green bar indicates the comfort percentages which not involve use of passive design techniques;
The deep green bar indicates the comfort percentages which involve use of passive design techniques.

设计说明

青海地区农牧民住房设计，以实际为出发点，充分考虑当地的自然环境与气候条件，从构思方案、规划布局到单体设计，主要通过被动太阳能利用技术，辅以主动太阳能利用技术及其他低技策略，来解决当地冬季集热采暖和夏季通风降温问题，并充分利用当地再生能源，建筑造价低，操作性强。

设计将太阳能利用最大化，改善当地冬季的室内环境，也不对环境产生污染，让当地居民喜迎每一天。

Design Explaination

The design of QingHai lower energy consumption residential house is found upon the analysis of natural environment and climatic condition, focusing on design proposal, layout plan and building design. We mainly use regional and feasible passive solar energy techniques, additionally use initiative solar energy and other low-tech method, so we could solve thermal-collecting in winter together with ventilation and cooling in summer. We make full use of local recycled resources, try to reduce build cost and simplify operation.

The design use a maximum of solar energy, try to make better interior circumstance, produce less pollution, let local inhabitant live joyfully.

Local Heating Condition

(1) The interior temperature is low. In winter, the temperature remains under 10°C even so many inhabitants concentrate in one bedroom to reduce heating consumption.

(2) High pollution. Unreasonable heat-collecting method and instrument together with use of numerous core wood and coal, bring about heavy inside and outside air pollution.

(3) Heavy economic burden. When we choose coal as main fuel, its consumption run up to 30~40 kg standard coal /(m²·a) - 1.5-2 times larger than city consumption, even the temperature is lower than city's. It's mainly for poor performance of heat preservation and inefficiency of heat-collecting system.

LOWER ENERGY CONSUMPTION OF RESIDENTIAL DESIGN

Layout Plan Design

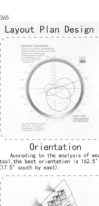

Orientation
According to the analysis of weather tool, the best orientation is 162.5° (17.5° south by east).

Layout
House comply with terrain, laying along contour direction, so we adjust the orientation to coordinate with terrain.

Winter Wind
Wind vane is NWbW in winter and the wall on northwest wide is heat-insulating wall.

Summer Wind
Wind vane is NWbN in summer and house layout has good guidance quality to wind.

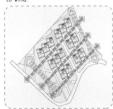

Road System
■ Village road ■ Group road ■ Footpath

Landscape System
■ visual corridor ■ landscape area

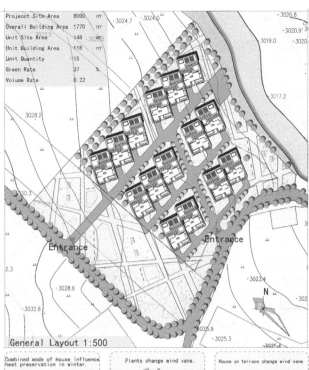

Project Site Area	8000 m²
Overall Building Area	1770 m²
Unit Site Area	148 m²
Unit Building Area	118 m²
Unit Quantity	15
Green Rate	37 %
Volume Rate	0.22

General Layout 1:500

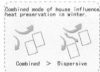
Combined mode of house influence heat preservation in winter.
Combined > Dispersive

Plants change wind vane.

House on terrace change wind vane.

Group Section Plan

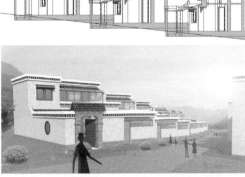

Houses are laying along terrace, we choose courtyard's mid axis to build that balance earth volume and save build cost.

生态藏居 Ecological Tibetan Dwelling
农牧民定居青海低能耗住房项目设计 2

LOWER ENERGY CONSUMPTION OF RESIDENTIAL DESIGN

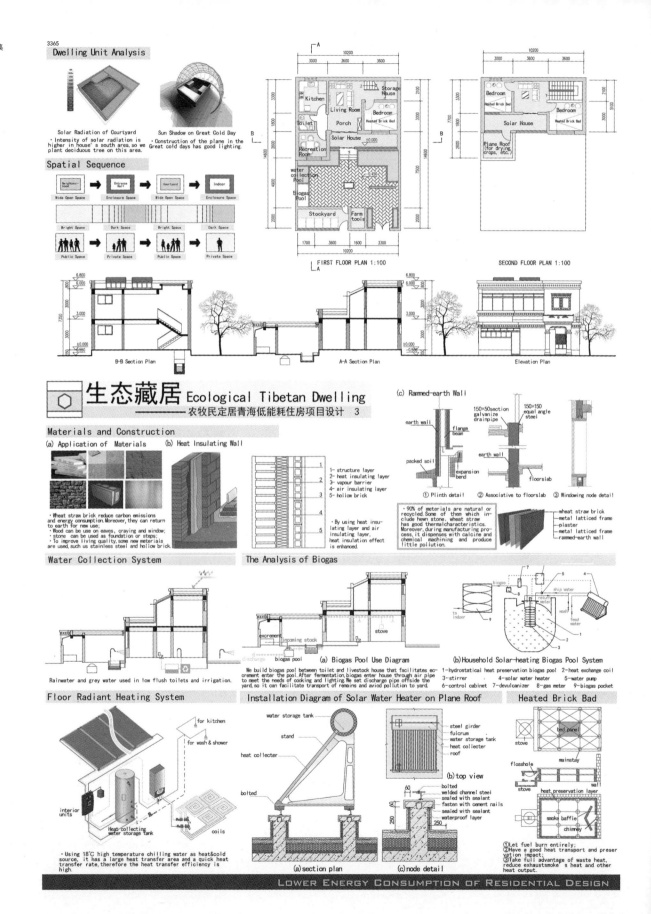

1 Ice-Storage Air Conditioning System

2 Ventilating layer Design in Overhead Insulating Roof

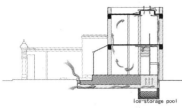
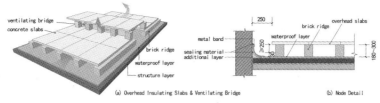

- As solar architecture's exterior construction's insulating fuction is better than traditional architecture, we choose ice-storage air conditioning—put ice into ice-storage pool and refrigerate them for summer use.

(a) Overhead Insulating Slabs & Ventilating Bridge (b) Node Detail

- We design ventilating bridge on roof, in one respect, using its surface to shelter against the sun, let the roof doing secondary transform, avoid solar radiant heat doing a direct action on exterior-protected construction; On the other respect, using wild pressure and heat pressure, especially natural ventilation, carry away amount of heat in ventilating bridge, thus reducing outdoors' thermal effect on internal surface.

3 Techniques' Comfort Range

4 Combined Passive and Active System

5 Ventilating Hole Design

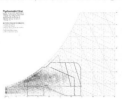
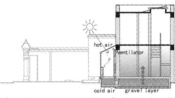
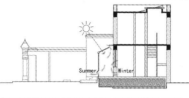

Design Techniques' Comfort Range on Psychrometric Chart

- On daytime, air temperature is higher after absorb solar radiant heat, the ventilator bubble hot air through gravel layer and heat the gravel, then air temperature will cool down; on night, cold air in room will back to solar house through gravel layer's heating.

- Open ventilating hole indoor and outdoor to improve winter & summer operating condition. (window or door can replace ventilating hole)

6 The relationship of adumbral devices and annual temperature

7 Shutter Insulation

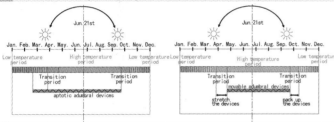
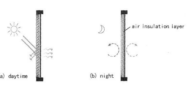

- Adjustable shutter on window help effectively control illumination and solar radiant heat on daytime, and effectively insulate by forming a closed temperature buffer layer.

8 Thermal Storage System

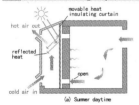

(a) Summer daytime (b) Summer night (c) Winter daytime (d) Winter night

生态藏居 Ecological Tibetan Dwelling
农牧民定居青海低能耗住房项目设计 4

9 Movable Shade Panels

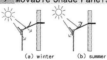

(a) winter (b) summer

- Shade panels can be turned around to control sunlight. The device has good view, small investment and is easy to control.

10 Insulating Glass

- Heat insulation glass can help achieve better results of sound & light control and heat insulation.

11 Arrangement of Plants

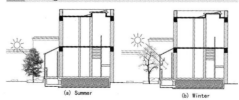

(a) Summer (b) Winter

- On summer, deciduous trees are a shelter from the sun when solar radiation is high; on winter, trees doesn't affect normal insolation after defoliation.

LOWER ENERGY CONSUMPTION OF RESIDENTIAL DESIGN

2015 台达杯国际太阳能建筑设计竞赛获奖作品集

综合奖 · 优秀奖
General Prize Awarded ·
Honorable Mention Prize

注 册 号：3367
项目名称：阳光庄窠（青海）
　　　　　Hold Sunshine (Qinghai)
作 　者：商选平
参赛单位：青海文旅投资有限公司

阳光庄窠
Hold Sunshine

01 概念 \ CONCEPT

Background

Location
The site is located in the Tuergan village, which is very close to the famous scenic—Qinghai lake.

Solar Situation
Qinghai Hehuang basin is a place with a lot solar energy resources. Because of the strong sunshine it is very suitable for solar technology.

Life Style
Local people are fond of "Shaiyangwa", which means "communicating in the sun".

Typical Building
The typical residence form is "Wall/House", which has a poetic Chinese name — "庄廓 (Zhuangkuo)".

SUNSHINE AIR THE PRIMARY SETTLEMENT
Brilliant sunshine and fresh air is not resources of our life, but a part of ourselves.

Design Concept

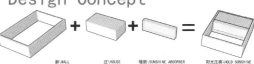

廓\WALL ＋ 庄\HOUSE ＋ 暖居\SUNSHINE ABSORBER ＝ 阳光庄廓\HOLD SUNSHINE

"庄窠"为青海河湟流域典型的民居形式，是当地气候环境和历史文化共同作用的智慧结晶。设计以"阳光庄窠"为概念，基于"庄窠"原型，通过提炼其应对气候的营建智慧、尊重原有地貌、融合新技术与本土工艺、集成适宜太阳能应用技术等策略，建构适宜环境、提升日常和精神生活品质、鼓励民众参与、具有模范性质的新农村生态聚落。

"Wall/House" is the typical residence form in Qinghai Hehuang basin. It demonstrates the local climate, history and culture. Our design is based on the concept of "Hold Sunsine", with the local "Wall/House" as prototype, which protects the original landscape and refines the building wisdom towards climate. On one hand, we combined innovated technology with traditional technology, and integrated appropriate solar technology. On the other hand, we created a demonstrative new rural ecological residence with comfortable environment, good mental living quality, and active public participation.

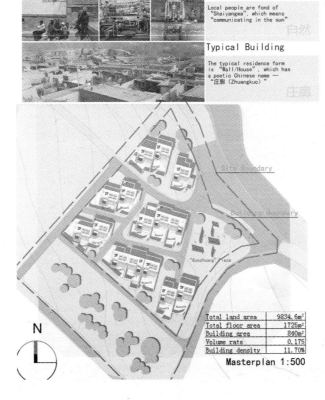

Total land area	9834.6m²
Total floor area	1725m²
Building area	840m²
Volume rate	0.175
Building density	11.70%

Masterplan 1:500

 Site
 Arrangement
 Rotate
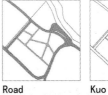 Road
 Kuo
 Zhuang

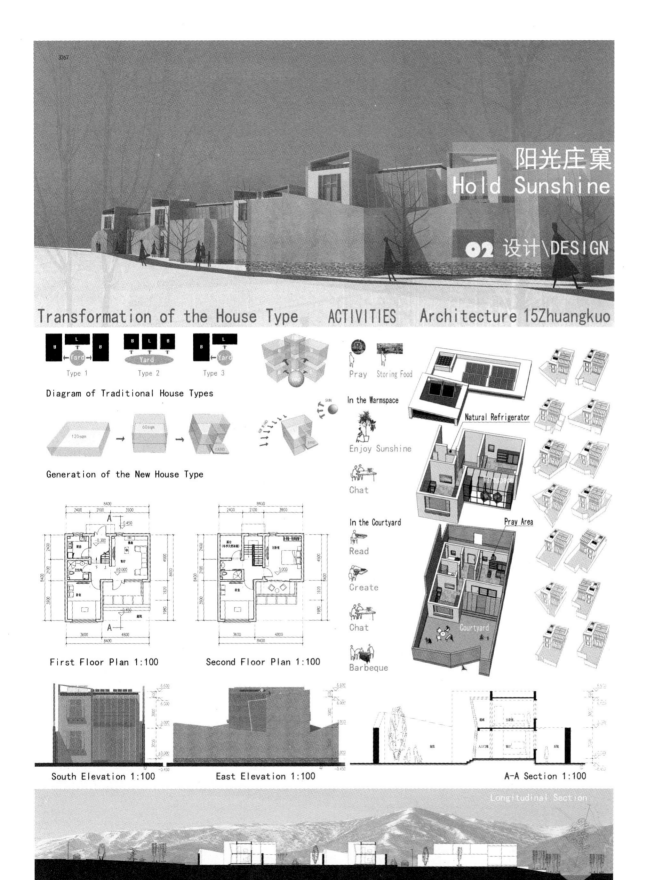

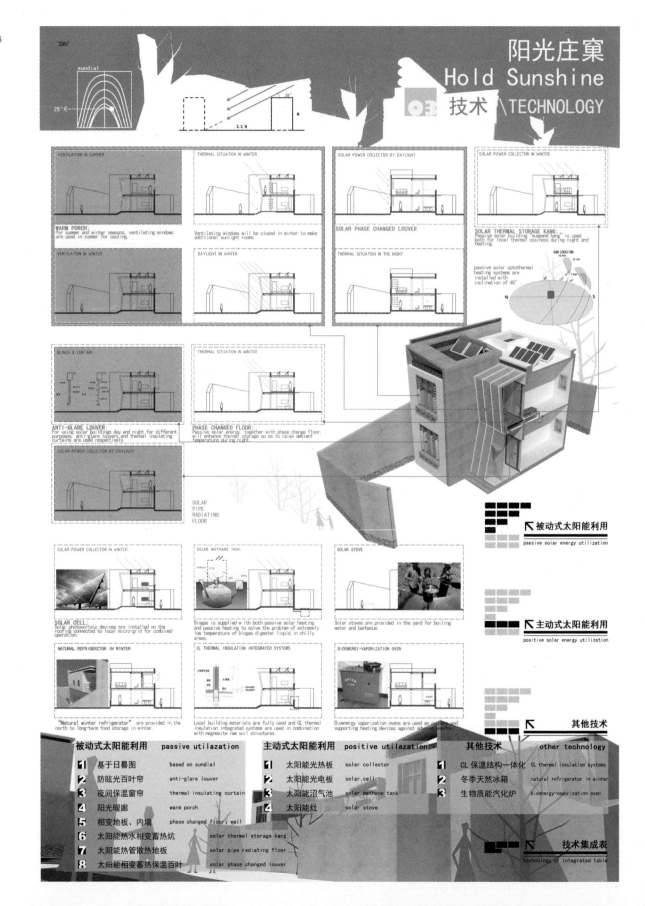

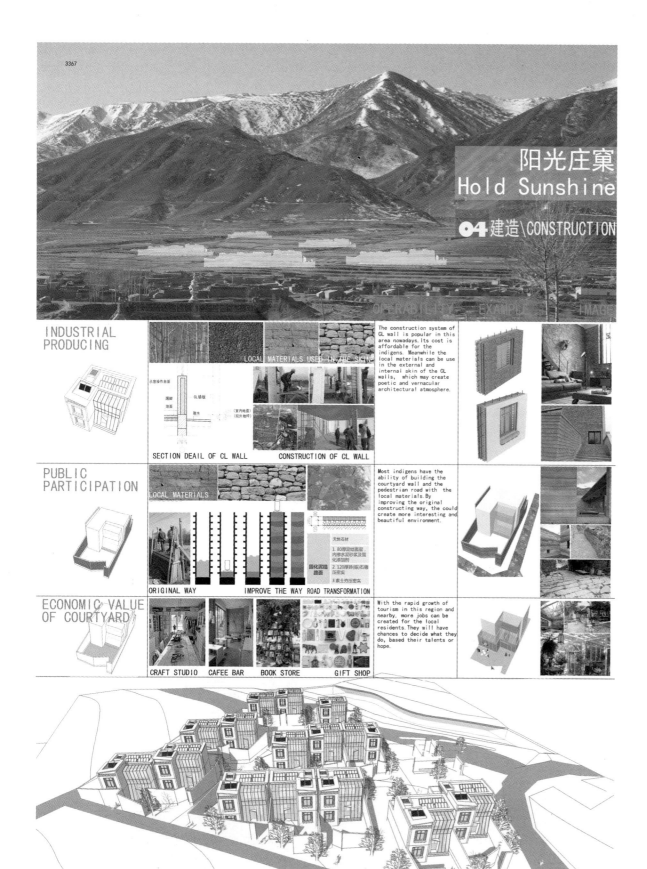

综合奖·优秀奖
General Prize Awarded · Honorable Mention Prize

注 册 号：3437
项目名称：土生土长（青海）
　　　　　Growth from the Loess
　　　　　(Qinghai)
作　　者：李和勇、刘荣伶、费怡巍、
　　　　　王竞竞
参赛单位：河北工业大学、天津大学

Husbandry is the traditional industry in the Tableland of Tibet and it is also the main and core industry for Qinghai province economic development. The basic industries that herders relies for survival.

Proportion of species in animal husbandry

羊 SHEEP 80.38%
牛 COW 17.58%
马 HORSE 1.04 %

Site analysis

Existing villages and roads to a relationship : vertical and parallel

Construction land　ranch　forest　village　stream　farmland

GROWTH FROM THE LOESS
—— 2015 台达杯国际太阳能建筑设计竞赛

设计说明：

　　本方案设计从当地传统民居的构造特点、空间结构、院落布局入手，结合对气候、光照、温度等的数据测算，对现有民居挑檐深度、屋顶形式、太阳房宽度和高度、院墙高度等细部尺寸、立面样式进行调整完善，设计出适应当地气候、来源地域文化的乡土住宅。又结合青海地区畜牧业发达的现状，在地块内布置配套羊舍、沼气池、雨水收集点等，将畜牧业、人、自然环境三者有机结合，形成自然资源有机循环；居住单体依据周边村落布局规律并结合最佳朝向和地块内具体自然风貌进行布局。

　　The project design started from the structure features, spatial structures and courtyard layouts of the traditional houses in the local area, and, according to data gathered on the climate, light, and temperature, adjusted and perfected some detail sizes and facade forms of the current folk houses including overhanging eave depth, roof form, sun room width and height, and courtyard wall height, in order to design local dwellings adapted to the local climate and original local culture. Combined with current situation of developed animal husbandry in Qinghai, supporting facilities including goat cots, methane-generating pits, rain water collection points were added, creating an organic combination of livestock, people and natural environment, to form an organic recycling of natural resources. The layout of living monomers is determined in accordance with the layout patterns of surrounding villages, combined with the optimal orientation and specific natural features within the site.

Material analysis　　　　　　**Analysis of local climate**

 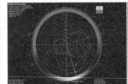 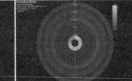 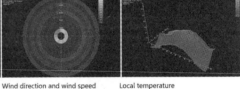

architecture form
the architectural form of Tibetan residence in Xining is Zhuangke　　details　　Best toward　　Wind direction and wind speed　　Local temperature

GROWTH FROM THE LOESS

—— 2015 台达杯国际太阳能建筑设计竞赛

Site plan 1:500

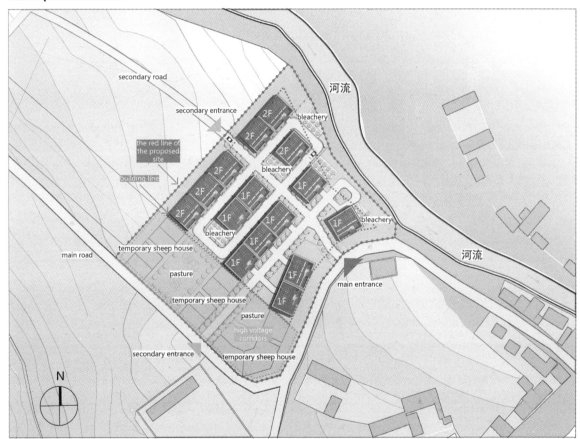

Local building temperature - time curve analysis

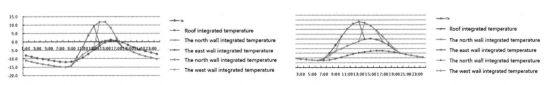

profile winter comprehensive temperature - time curve   profile summer comprehensive temperature - time curve

GROWTH FROM THE LOESS

2015台达杯国际太阳能建筑设计竞赛

Local building research

Anaylsis of plan form

"Zhuangke" is the local traditional houses, which is a regular quadrangle layout. The principal room is always builded with pine poles, beams, purlins as load-bearing structure. And it is made by retaining wall of soil and decorated with fine wood carving. The form of principal room mainly includes "font", "key" and "tiger head on her head".

1. lobby
2. bedroom
3. kitchen
4. storeroom
5. utility room
6. toilet
7. livestock room
8. fodder room
9. entrance
10. tools room
11. courtyard
12. scripture hall

The Typical Plane Form

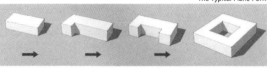

The Form of "—"

The Form of "⌐"

The Form of "⊐"

Anaylsis of section form

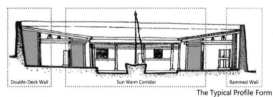

The Typical Profile Form

(1) **Rammed wall**
The rammed wall is build with the use of rich local loess or adobe masonry, and it is tall (5m) and thick (0.8~1.2m) wall; loess (adobe bricks) has small thermal conductivity and big heat capacity.

(2) **Sun Warm Corridor**
The sun warm corridor can store the heat of sun in the daytime of winter, and release to the indoor with lower temperature in the night. So it improves indoor comfort level.

(3) **Double-Deck Wall**
The double wall is composed of rammed earth walls and houses load-bearing wall, and it forms an air gap between them. So it has a very good effect on heat insulation of house.

Extraction of plan prototype

First Plan / Second Plan

- kitchen
- bedroom
- living room
- bathroom

Idea (Annular Air Layer)

Compound analysis

The Sunshine Analysis of Houses with 150㎡ Homestead

8m=4.2m (room depth) +1.8m (sunroom) +3m (garden)
Through the above analysis, only if the south wall height is less than or equal to 4.5m, it can ensure the yard lighting.

Solar house analysis

south depth of the solar house

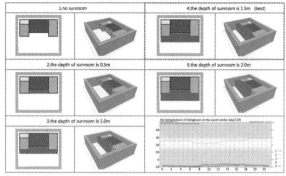

Conclusion
(1) After the appearance of sun warm corridor, the temperature of livingroom is significantly increased.
(2) With the increasing wide of south sun room, the temperature of livingroom is not widely changed.

East depth of the solar house

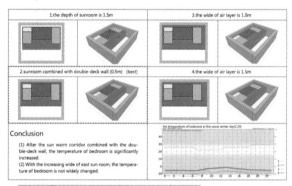

Conclusion
(1) After the sun warm corridor combined with the double-deck wall, the temperature of bedroom is significantly increased.
(2) With the increasing wide of east sun room, the temperature of bedroom is not widely changed.

Roof overhangs analysis

Roof form

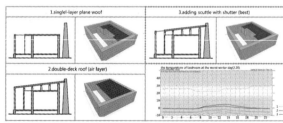

Conclusion
(1) After adding the double-deck roof, the temperature of bedroom is significantly increased.
(2) After adding the scuttle with shutter, the temperature of livingroom is increased littlely on the noon.

Overhangs form

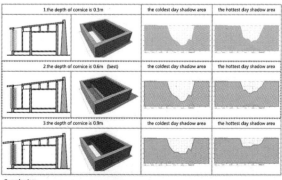

Conclusion
(1) The depth of cornice ≤0.6m, the shadow area of this wall is not widely changed.
(2) The depth of cornice ≥0.6m, the shadow area of this wall is significantly increased on the coldest day.
(3) With the increasing depth of cornice, the shadow area of this wall is significantly increased on the hottest day.

GROWTH FROM THE LOESS

—— 2015 台达杯国际太阳能建筑设计竞赛

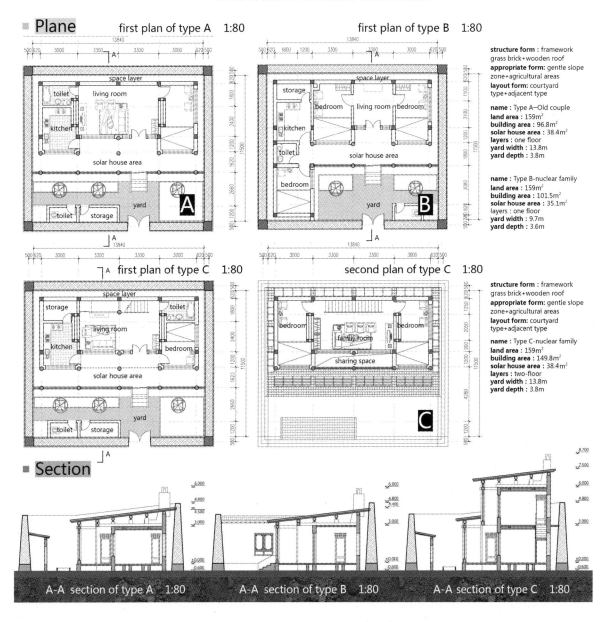

Solar house analysis

Chimney analysis
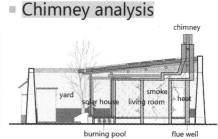

GROWTH FROM THE LOESS
—— 2015 台达杯国际太阳能建筑设计竞赛

Building analysis
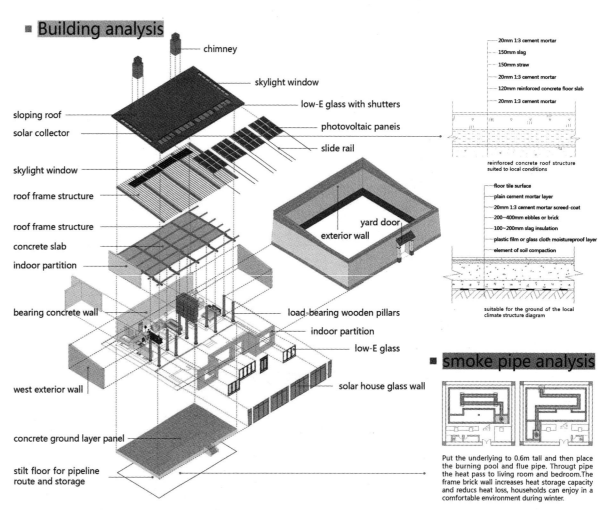

smoke pipe analysis

Put the underlying to 0.6m tall and then place the burning pool and flue pipe. Througt pipe the heat pass to living room and bedroom. The frame brick wall increases heat storage capacity and reducs heat loss, households can enjoy in a comfortable environment during winter.

Trombe wall system analysis
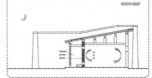   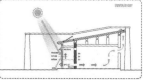

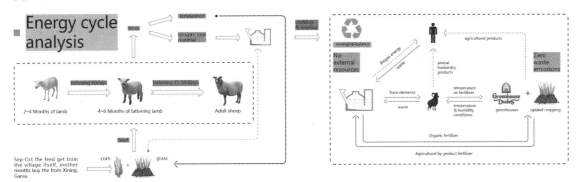

GROWTH FROM THE LOESS

—— 2015 台达杯国际太阳能建筑设计竞赛

综合奖·优秀奖
General Prize Awarded · Honorable Mention Prize

注　册　号：3456
项目名称：阳光与美丽的交"措"（青海）
　　　　　Sunshine Embrace Beauty
　　　　　(Qinghai)
作　　者：艾洪祥、王　洋、张天宇、
　　　　　郭冰月、赵珍仪、范　斌、
　　　　　陈　强、吕高标、宋　薇、
　　　　　魏宏毫
参赛单位：山东建筑大学、重庆大学

阳光与美丽的交"措"
Sunshine Embrace Beauty

"Sunshine" refers to the use of solar energy and the relationship between light and shadow; "Beautiful" refers to the activities of space, landscape (viewing platform) and the village itself; "Jiao cuo" is a design strategy, can improve efficiency and produce the form beauty measures, at the same time the "measures" in the Tibetan language means lake, reminiscent of Tibetan culture.

设计说明： 本方案从阳光与美丽乡村的主题出发，通过对场地气候、地形地貌、人文环境等进行充分调研分析，以太阳能的利用与公共活动空间的辩证关系为切入点，提出"交措"的设计理念和方法。并将其贯彻到组团规划、建筑单体以及太阳能利用技术的设计之中。形成活动空间的营造、景观视线的布置、区域能源、建筑造型、建筑室内环境的调节等方面的设计策略。

Design description: This scheme start from "The sunshine and beautiful country", based on the analysis of the site climate, topography, the humanities environment etc.. Take the dialectical relationship between the use of solar energy and public space as the breakthrough point, put forward "the design concept and method of Interleaving". And to carry out the planning group, single building and solar energy utilization technology in design, the formation of the activity space design, landscape sight layout, regional energy, the architectural style, building indoor environment regulation and other aspects of the design strategy.

设计前分析 Pre analysis

Temperature contours — Belongs to the building climate divisions in China cold area B district. Here, buildings do not need heating from may to September, and need heating during other month.

Qinghai Xining Tuergan Prevailing wind direction — The dominant wind direction in winter is the northwest by west, average wind speed 1.75 m/s; Dominant wind direction in summer is north by northwest, the average wind speed 1.7 m/s.

Bioclimatograph — By mapping bioclimatic weather chart to determine the suitable climate response which required to create thermal comfort in this climatic conditions.

Solar Analysis — Solar radiation resources is rich, annual average sunshine hours are 2718.6 hours. The best orientation is 25° south by east.

Traffic analysis — Site surrounded by the village road, the village roads extend from 109 national road to other villages.

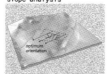

The direction of the slope analysis — North and south elevation difference is nearly 5m. Isoheight and the best building orientation are parallel building of the field, it provid the opportunity for building to get the best orientation and decrease earth volume during construction.

Topography analysis — Outside the venue near a substation, the north is a brook; In the field, a high tension line corridor to separate construction land and the village road.

Energy analysis — Outside the venue near a substation, the north is a brook; In the field, a high tension line corridor to separate construction land and the village road. About the unfavorable factors, the high voltage corridor can be an energy center about road land and substation to provide residential electricity.

Photoes of Current Situation

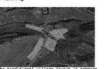

Planning — The traditional village layout is compact, conform to the topography, residences of adjacent Zhuangke usually share walls, in zonal distribution. Villages have public space.

Building — Local traditional architectures are raw soil building—Zhuangke. Features: exterior wall cover apart wall; rejecting outside room; internal cohesion; Flat roof; block feeeling is obvious.

Colour — After analyzing the Tibetan and local color, extract combinations which can reflect the local natural and humanistic color.

What the Villagers Said — The main production way of local villagers is agriculture and animal husbandry, believe in Tibetan Buddhism. They hope to have a roof terrace to dry grain in the sun, an oratory and enough sunshine to keep warm.

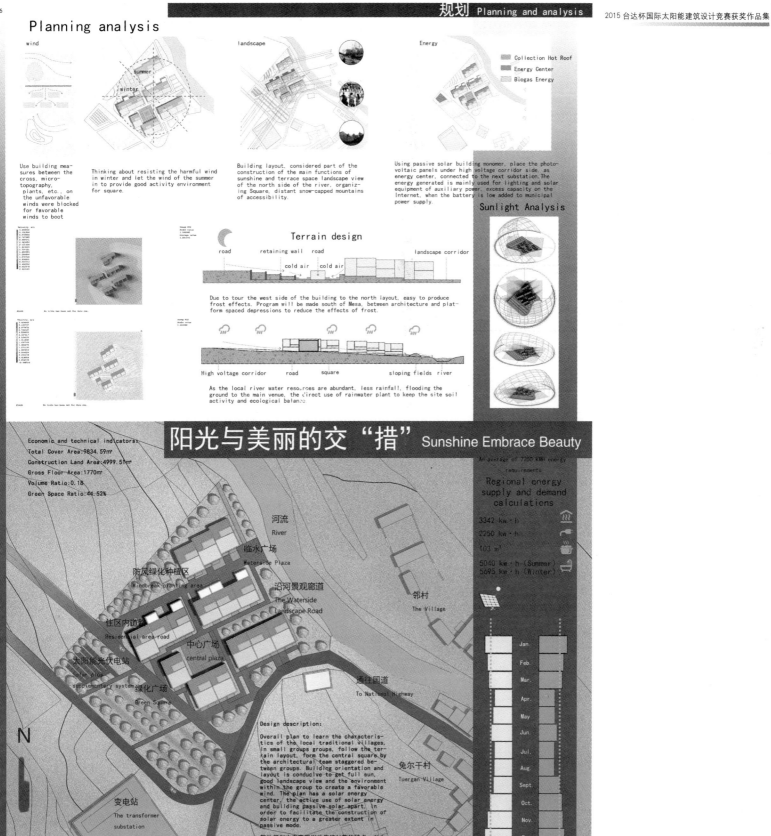

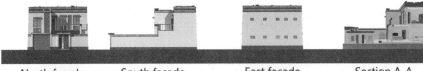

建筑设计 Architectural design

Scene view

Elevator pull food

Balcony family reunion

Terrace viewing

Elevation and section

North facade　South facade　East facade　Section A-A

Shape Analysis: Learn traditional architectural building blocks of the body and building a sense of exclusion external interface, opening up features.

Color Analysis: Traditional Tibetan color language: terrace mounds of using color to emphasize the balcony of the family public space.

First floor plan 1:75　　Second floor plan 1:75

✤ Heating intensity

Inside the temple worship

Residential restaurant interior

阳光与美丽的交"措"　Sunshine Embrace Beauty

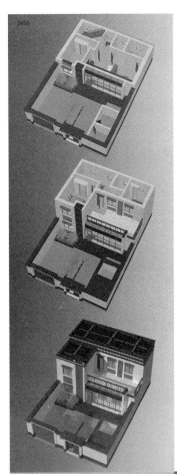

模拟与策略 Simulation and strategy

Shadow area

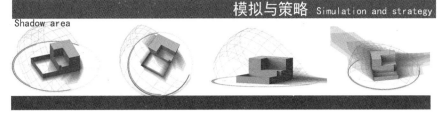

6.21 Shadow area   6.21 Shadow area   12.21 Shadow area   12.21 Shadow area

Lighting simulation   Wind simulation

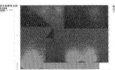   

first floor daylight factor   second floor daylight factor   First floor Wind simulation

first floor illuminance   second floor illuminance   Second floor Wind simulation

Architectural shadow simulation

The design separate two bedroom on second floor, form a shadow buffer as a place for flat roof, family party and viewing. solve the conflicts on sunlight of bedroom and the space of flat roof.

阳光与美丽的交"措" Sunshine Embrace Beauty

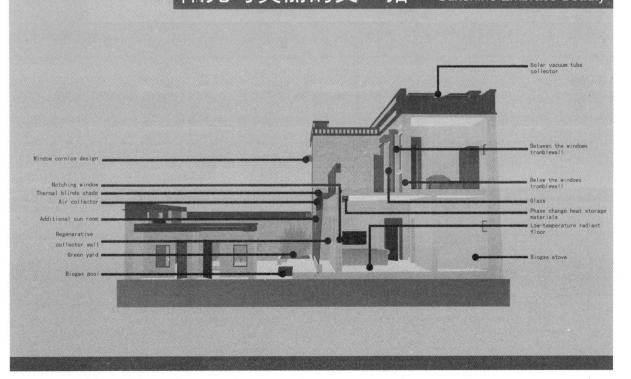

Labels: Window cornice design; Notching window; Thermal blinds shade; Air collector; Additional sun room; Regenerative collector wall; Green yard; Biogas pool; Solar vacuum tube collector; Between the windows tromblewall; Below the windows tromblewall; Glass; Phase change heat storage materials; Low-temperature radiant floor; Biogas stove

技术分析 Technical analysis

Solar heating and hot water supply system

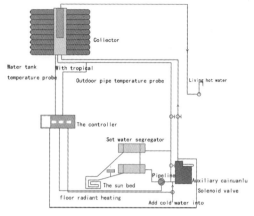

The system uses solar vacuum tube collectors to collect heat of solar radiation, and combined with auxiliary energy to satisfied the requirements of heating and domestic hot water. The Heating system uses water as medium to heat indoor by low temperature floor radiation, when we use the hot water in our daily life, the tap water heated by heat exchanger in water tank and then we canuse it by adjusting Water mixing valve.

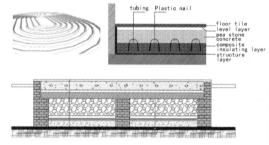

- Cement mortar screed-coat
- Packed bed of fine stone concrete
- Solar water heating coil
- Aluminum foil reflector
- Polystyrene insulation layer
- Precast concrete panel
- Pebble bed

- Pebble bed
- Fine stone concrete protective layer
- Coiled material waterproof layer
- Cement mortar screed-coat
- Concrete pad
- Concrete leveling layer
- Element of soil compaction

Biogas utilization

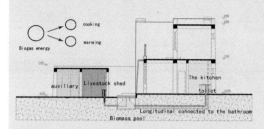

Biomass marsh gas pool set up near the side of the kitchen, toilet and live-stock shed, it discharge the waste to the Biomass pool to produce methane for life and the auxiliary heat source of heating, the marsh gas pool uses biomass for heat insulation, to ensure the methane supply.

Window cornice design

Sunblind absorb the form of the traditional Tibetan sunblind, in order not to affect the setting sun effect, form vents.

Notching window

Use cut processing on the Windows inside corner to increase the receiving surface of solar radiation of South window.

the window on thr south of badroom set the Heat insulating board in able to keep the temperature at night.

Solar Kang

Open the heat preservation plate during the day, black pebbles save heat.

Closed insulation board at night, the pebbles release heat continuously, so it can heat the bed together with hot water floor

Set the sun bed In the old man's room, to adapt to the old man's habits and customs, the sun bed is heated by solar energy water heater floor radiation and black pebbles, solve the flaws of traditional heated brick bed, such as diffi-cult to adjust temperature, low thermal efficiency, environmentalpollution, poor thermalcomfort and so on.

阳光与美丽的交"措" Sunshine Embrace Beauty

Additional sunlight room

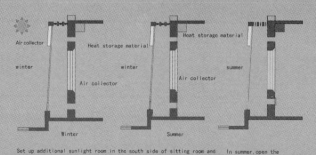

Set up additional sunlight room in the south side of sitting room and dining-room, use hot air circulation convection and thermal radiation heat the indoor air at daytime,the air heat collector towards the sun will transfer the heat efficiently to heat the heat storage material in the sunlight room, release heat for room at night.

In summer, open the shutter, to Strengthen the ventilation.

Ventilation

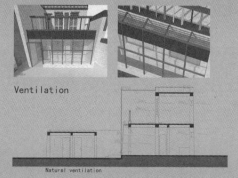

This region is given priority to with the north-northwest winds at summer, the building use the natural ventilation can achieve good ventilation effect.

技术分析 Technical analysis

Solar chimney

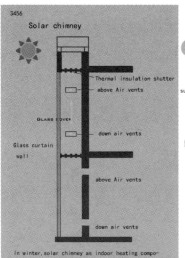

In winter, solar chimney as indoor heating components heat the bedroom and sitting room of the secondfloor heating, the implementation method is closed the insulation shutter of the solar chimney.

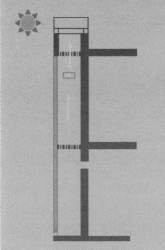

In summer, open the thermal insulation shutter of the sunchimney at summer, Rise Chimney wall temperature and heat the air air flow.

The external retaining structure

Tromble wall

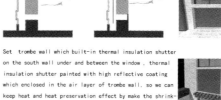

Set thermal shutter in bottom vent of trombe wall at summer, took the hot air to the outside.

Set trombe wall which built-in thermal insulation shutter on the south wall under and between the window, thermal insulation shutter painted with high reflective coating which enclosed in the air layer of trombe wall, so we can keep heat and heat preservation effect by make the shrinkage of thermal insulation shutter and open of vent in the wall.

North window size determination

In order to reduce energy loss, reduce the North window as far as possible when of the daylighting satisfy the requirement. The daylighting simulation analysis, draws the window daylighting coefficient of transverse dimensions longer relatively high conclusion.

North window size determination

Choice of white glass single silver low-E glass insulating film as a window glass.

Shading   Plant shade

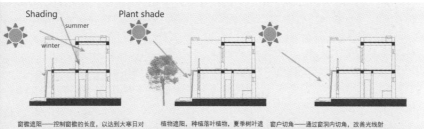

窗檐遮阳——控制窗檐的长度,以达到大寒日对阳光无遮拦,夏至日可以全部遮挡住阳光
Sunblind set sun - control the length of the sunblind, achieve sunshine during Great Cold, and cover sunshine during the summer solstice.

植物遮阳,种植落叶植物,夏季树叶遮阳,冬季树叶掉落
Plants shade, planting deciduous plants, use leaves as shade at summer, leaves fall at winter.

窗户切角——通过窗洞内切角,改善光线射入室内的均匀度,减少眩光,改善光环境
cut the window corner-use the inscribe angle hole of the window to improve the homogeneous degree of the light riping into indoor room to reduce glare to improve the light environment.

阳光与美丽的交"措" Sunshine Embrace Beauty

External structure in addition to the south wall using aerated concrete, the other part is the construction of EPS module. EPS module prices down disposable mouth pouring plug and concrete, insulation and building structures achieve the same life cycle.

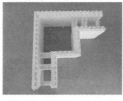 

South to the wall using aerated concrete beneficial use of solar radiation wall. EPS module thermal conductivity is generally 0.03W/(m·K), can achieve good insulation effect.

Exterior wall decorative surface layer structure.

综合奖·优秀奖
General Prize Awarded · Honorable Mention Prize

注 册 号：3571
项目名称：光弧（青海）
　　　　　Solar Arc (Qinghai)
作　　者：李惠桦、晏凌峰、杜杨、
　　　　　周清华、吴少博、薛超、
　　　　　刘建松
参赛单位：北京交通大学

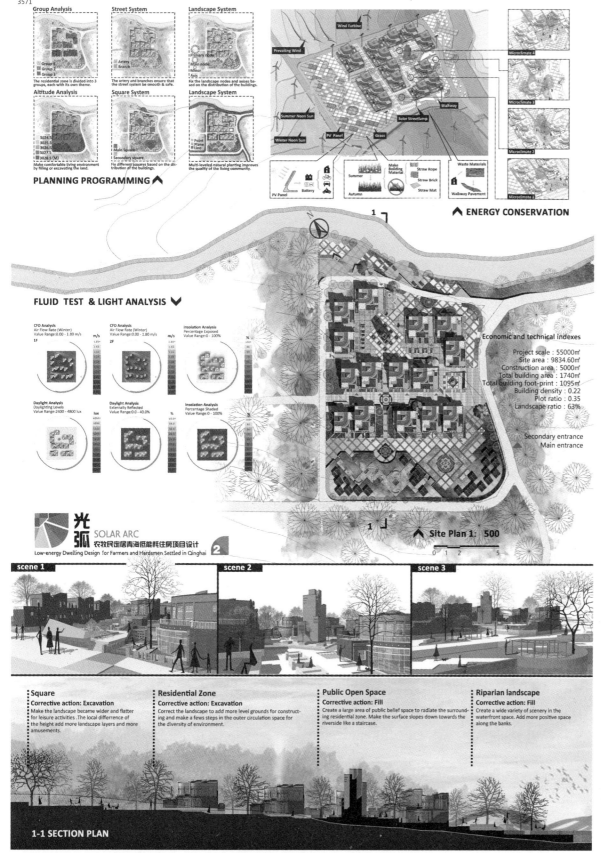

3571

Concept Generation

Concept is derived from the growth of plants. During the day, plants absorb the sunlight to provide heat for themselves. At the same time, consumption part of its nutrients work as an auxiliary heat source. At night, plants mainly rely on the self consumption with heat. Simulating the characteristic of plants, we design a sun space to absorb the daylight and use rural traditional heating measure like stove, fire wall and fire kang to form auxiliary heat source. Then put the function rooms between the two heat sources, achieving a good thermal environment.

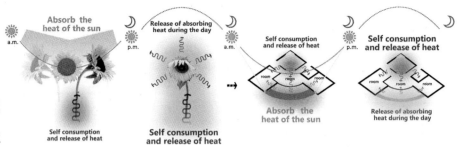

1 Form a terrace 2 Daily sun path 3 Facing south 4 Rain water collection & expand the sunny 5 Solar energy utilization 6 Cooking stove, fire wall & chimney 7 Courtyard & barn

光弧 SOLAR ARC
农牧民定居青海低能耗住房项目设计
Low-energy Dwelling Design for Farmers and Herdsmen Settled in Qinghai

▲ Form Generation
The sun's orbit is the most important design basis, in order to maximize the absorption of sunlight, the sun space is set to an arc

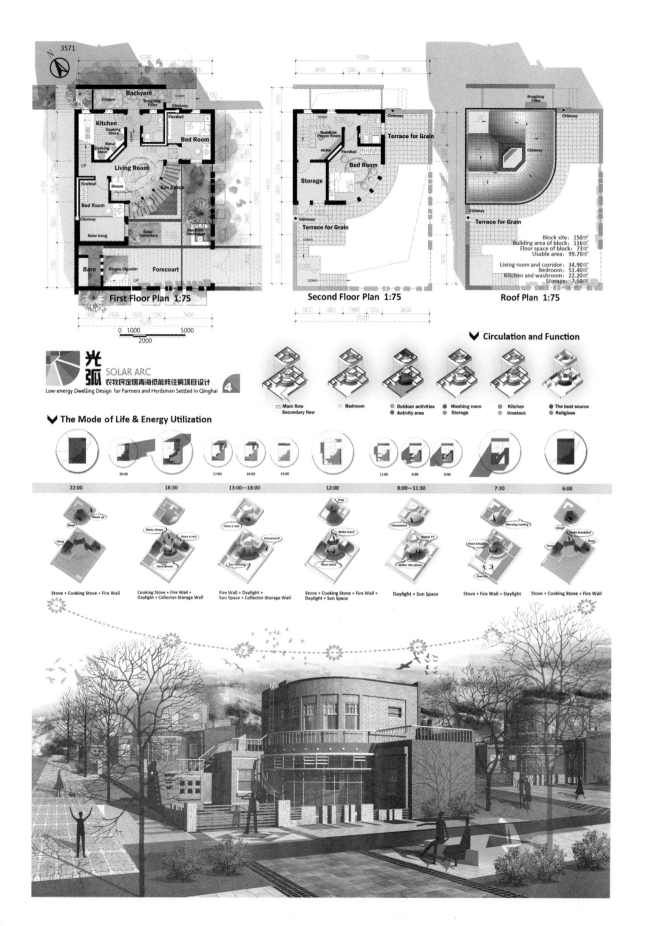

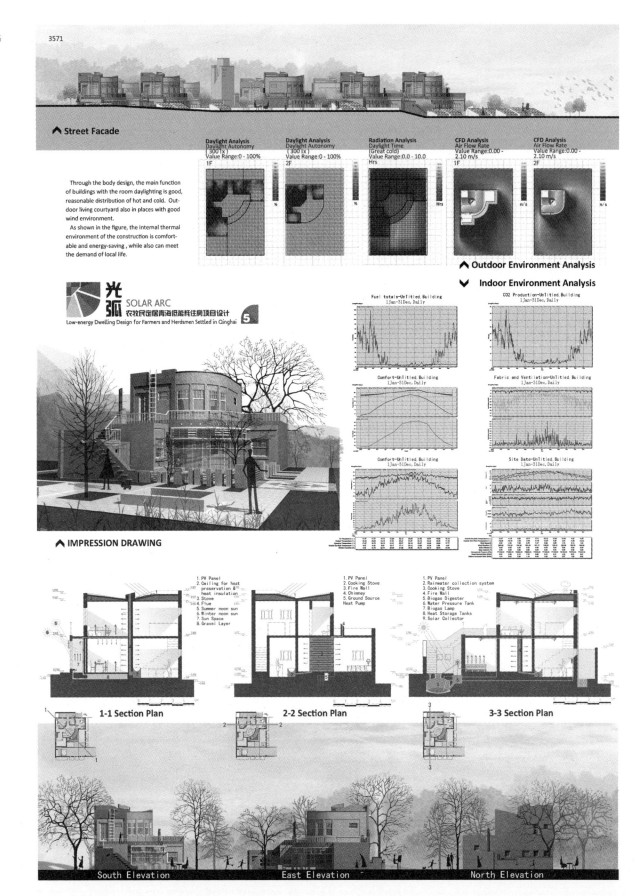

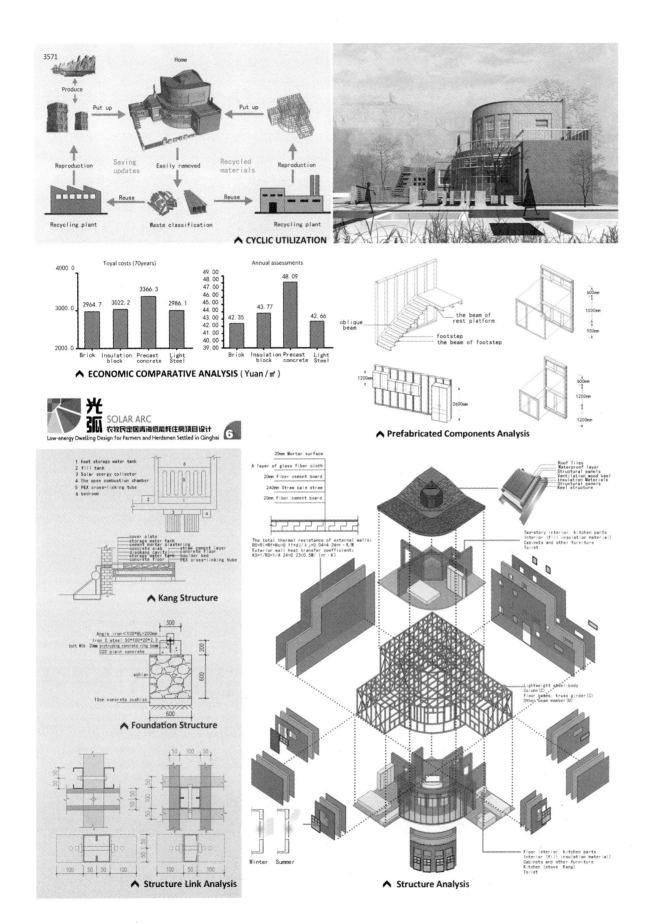

综合奖·优秀奖
General Prize Awarded · Honorable Mention Prize

注　册　号：3626
项目名称："牧"光城——沐光·暮光·融光（青海）
The City of Pasture & Sunshine (Qinghai)
作　　　者：刘冲、姜羽平、胡春霞、梁一航、倪翰聪、杨晶晶、毕文蓓、张习龙
参赛单位：西安建筑科技大学

3626 "牧"光城——沐光·暮光·融光
The City of Pasture & Sunshine
农牧民定居青海低能耗住宅设计

BACKGROUND ANALYSIS
CITY IMAGE ANALYSIS

GEOGRAPHY AND CLIMATE ANALYSIS

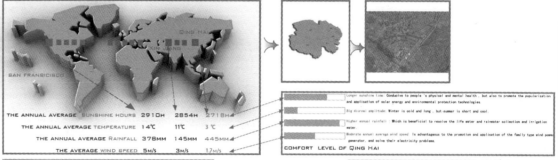

CURRENT SITUATION ANALYSIS
THE CURRENT SITUATION OF TUERGAN VILLAGE

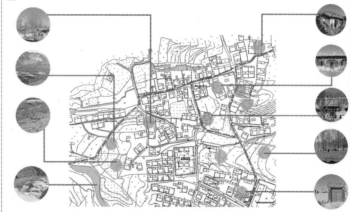

PRE-PROGRAM

本方案以理性逻辑作为设计思路，在分析场地和建筑现状、青海气候特点、牧民行为模式及需求的基础上，凝练青海地区的传统建筑风格并进行创新，通过阳光间、太阳能光伏电板、草砖、双层中空Low-E玻璃、节能热炕等技术来解决当地冬季漫长寒冷的问题，同时，利用自然通风、绿化墙等来解决夏季炎热的问题，另外，雨水收集系统、人工湿地净化生活污水系统以及风力发电系统的运用，成为了方案技术层面的主要特色。

以"牧"光城为名，一是指方案的定位为农牧民定居住宅设计；二是指在建筑材料的选用上，选用当地的木材作为阳光间及其他房间的主要材质；三是指在暮光城之下，牧民安居乐业的美好图景。因此，牧民、木材、暮光三者相互作用，共同塑造和激活牧民对美好生活的向往，让建筑、人、景共生。

THE CURRENT SITUATION OF SITE AND BUILDINGS

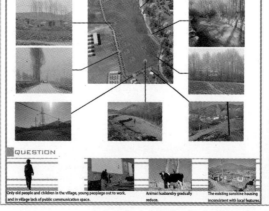

SITE ENVIRONMENT ANALYSIS

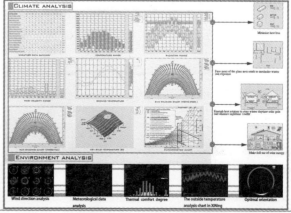

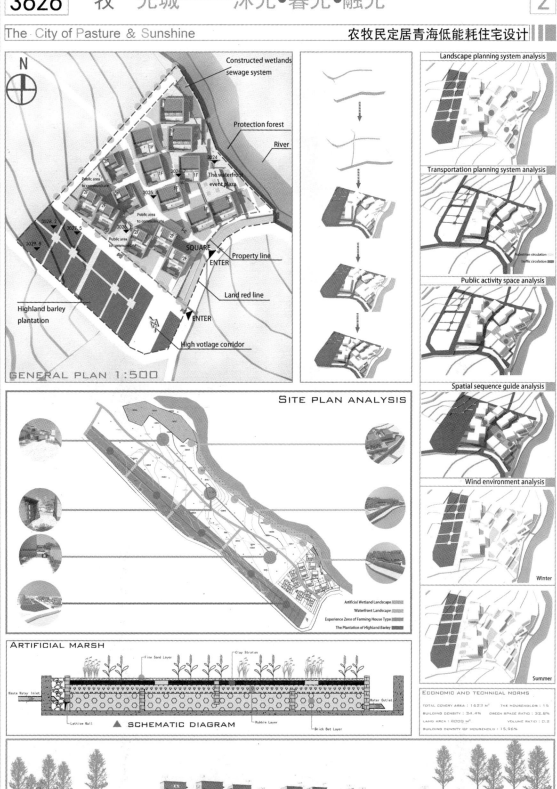

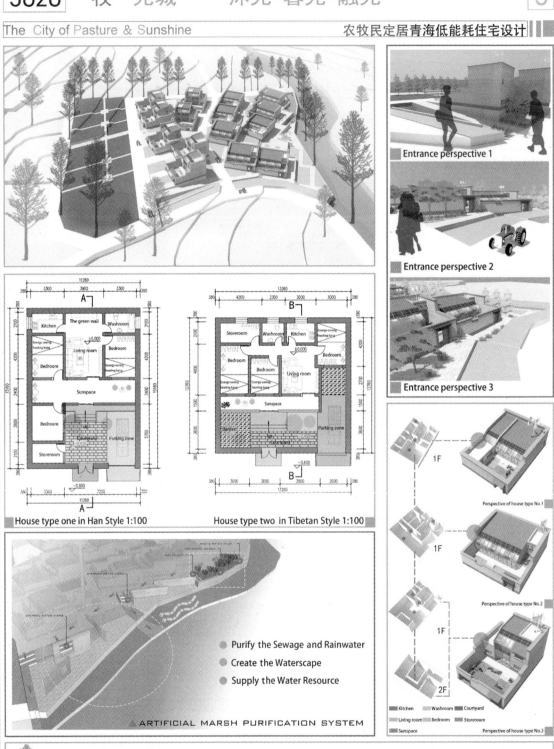

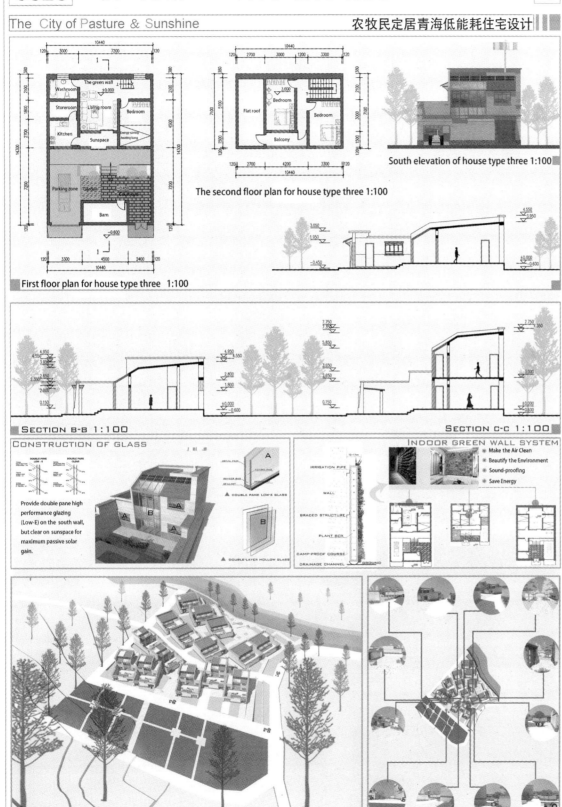

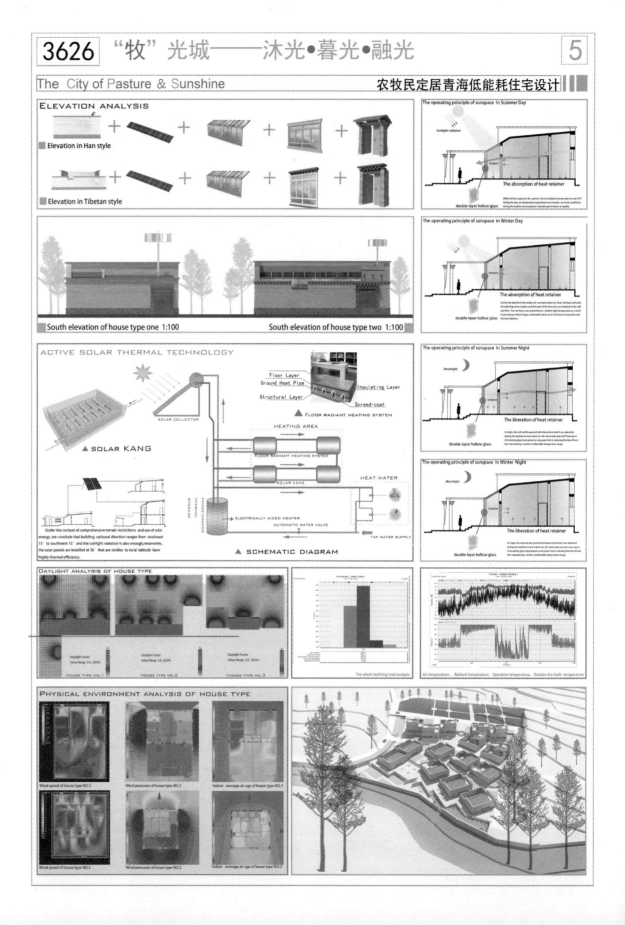

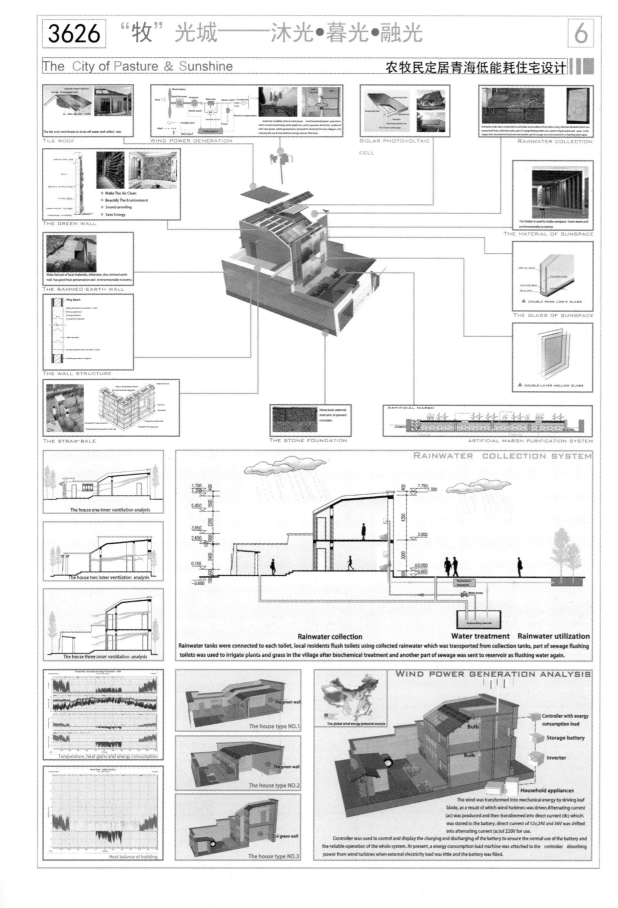

综合奖·优秀奖
General Prize Awarded · Honorable Mention Prize

注 册 号：3638
项目名称："1/2阳光"安多哇自宅（青海）
　　　　　"A Half of Sunshine"
　　　　　AnDuoWa Home (Qinghai)
作　　者：王轶楠、王冠宇、叶兆丹、
　　　　　闵韵然
参赛单位：重庆大学

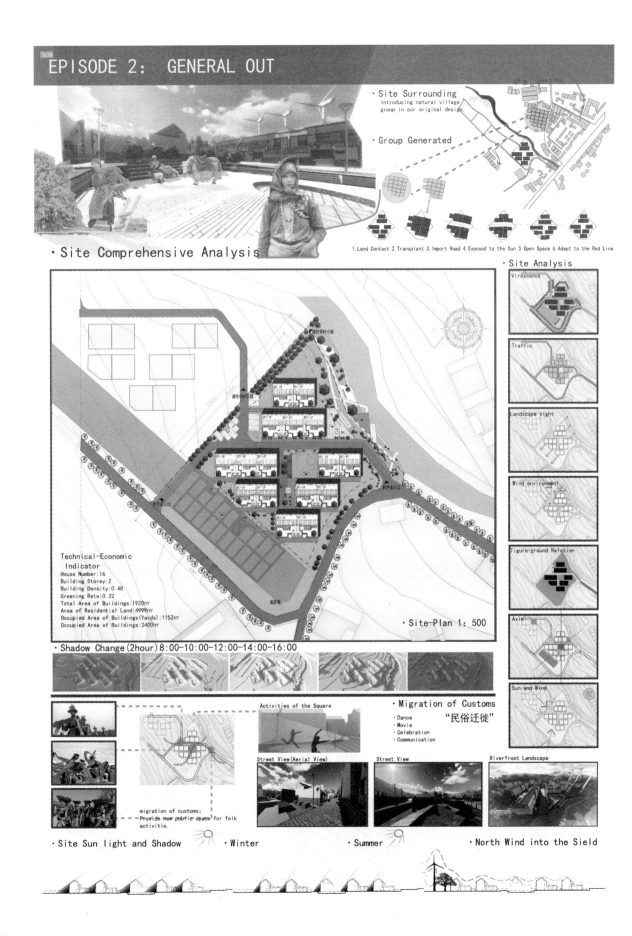

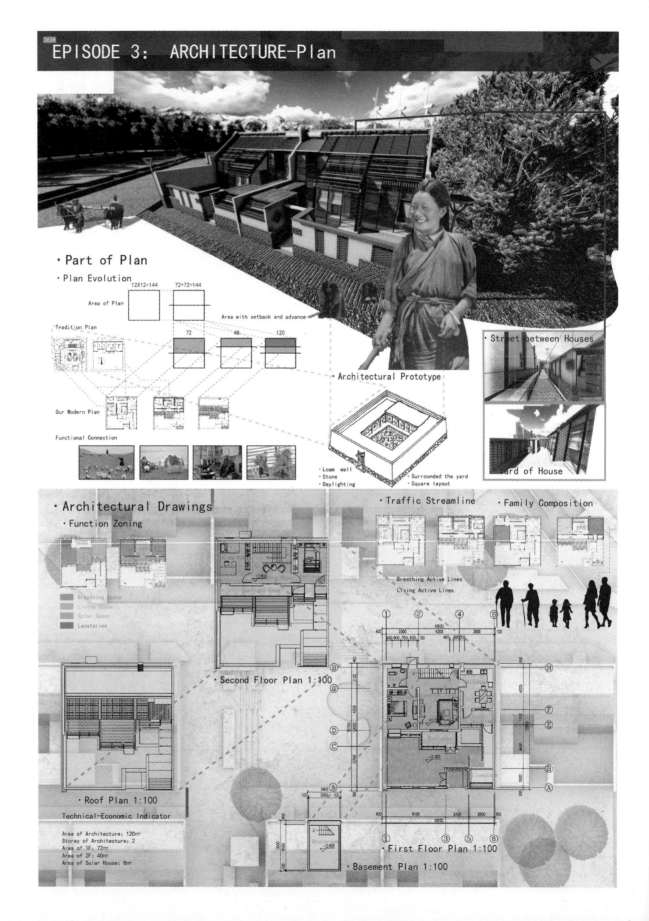

EPISODE 4: ARCHITECTURE-Stereoscopic

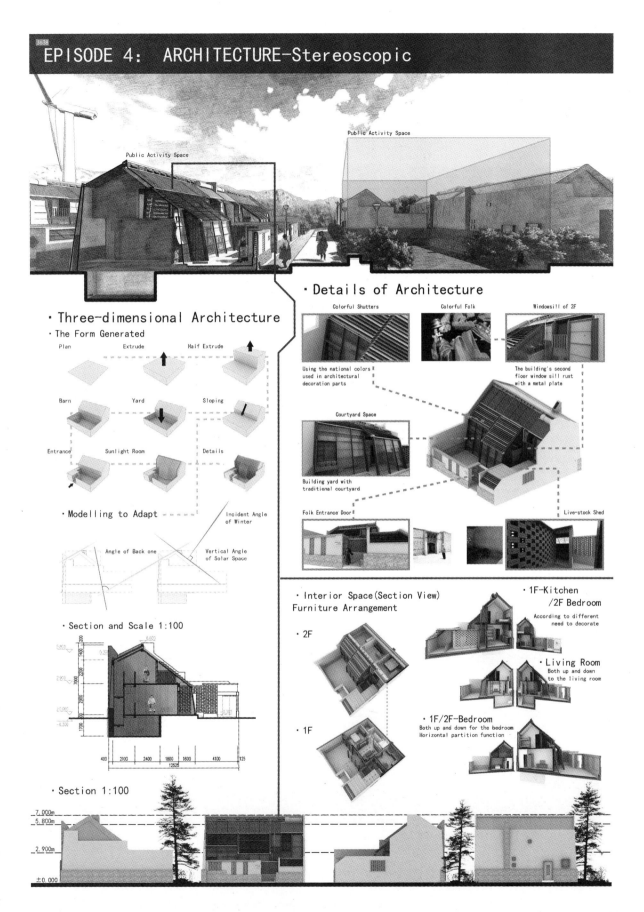

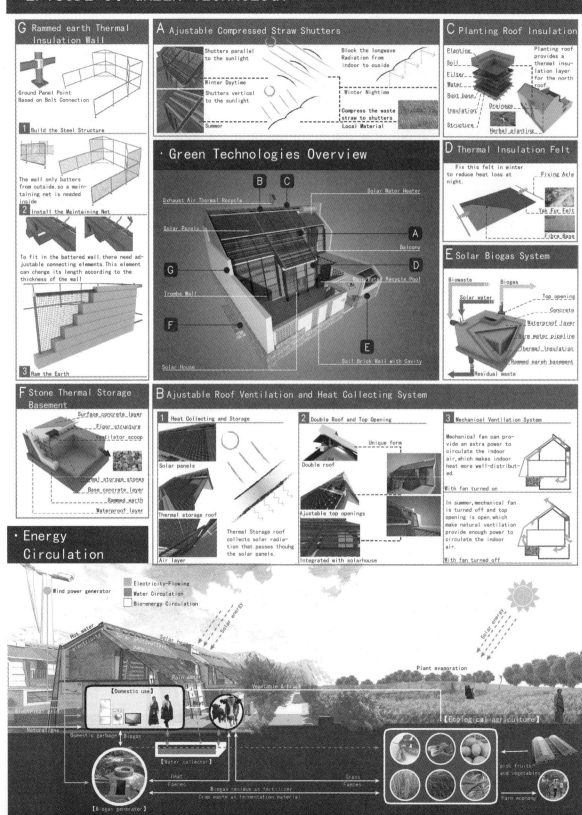

EPISODE 6: OPERATION ANALYSIS

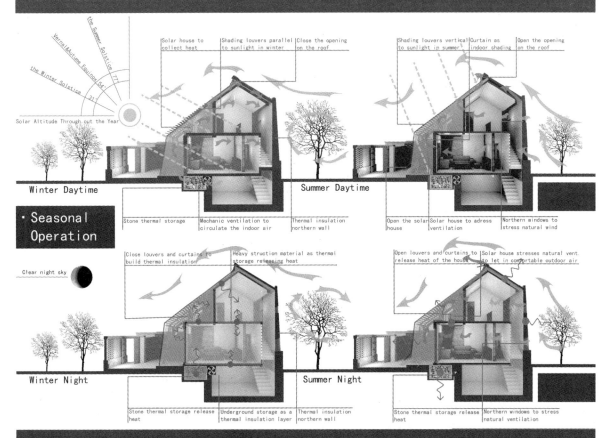

· Seasonal Operation

· Digital Simulation

1. Solar House Simulaition

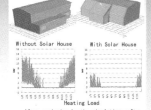

After analyzing by the simulating software DEST, we can find out the all-year-accelerated heating load decreses from 98.67kWh/m² to 44.99kWh/m² thanks to the solar house, just 54.4% of the heating load of the building without solar house.

2. Solar Panel Simulation

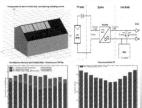

The Amount of the Generated Electricity

According to the simulation of PVsyst, one single house can generate 1078kWh/a electricity, making up to more than 90% of requried electricity (118kWh/a according to the yearbook of Qinghai Province).

3. Solar Water Heater

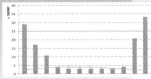

■ Heating Amount
■ Domestic Hot Water Amount
▲ Solar Hot Water Produced Amount

From the gragh, we can see that the hot water produced by the solar heat collector is sufficient for domestic use through the whole year only except for a few months in winter in which the hot water is a little insufficient. However if we need the hot water heating system, an extra boiler or heat pump must be installed.

4. Site Wind Circumstance

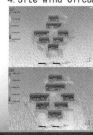

Using the CFD software Ansys, we have simulated the sie wind circumstance in summer and winter. Selecting the 1.5m-above-ground plane as the researching plane, we find out that the north-western part of the site has the strongest wind of which the velocity is 2m/s. So we plan to plant some windbreak trees in that part of the site.

5. Summer Indoor Ventilation

In summer, in order to creat indoor ventilation, we need to open the window of the north wall, the solar house, and of the roof. We have simulated the indoor natural ventilation by the CFD software Ansys.

Natural wind comes in through the northern window. After going through the main part of the house, it finally reaches the solar house. Being heated by rhe solar energy, hot air goes upward to the roof ang is exhausted out of the house.

Because the unique form of the roof, there has been created a space of subatmospheric pressure in front of the top opening on the roof, thus avoiding the mutual offset between heat and wind pressure.

6. Shading Simulation

After analyzing by Ecotect, we can find that over 90% of the site has more than 2-hour sunshine in the winter solstice. Furthermore, there is no mutual shading between the house of this site.

综合奖·优秀奖
General Prize Awarded · Honorable Mention Prize

注 册 号：3768
项目名称：辞柯（青海）
　　　　　CiKe (Qinghai)
作　　者：安　娜、魏宏毫、张晨悦、
　　　　　提姗姗、邢龙飞
参赛单位：山东建筑大学

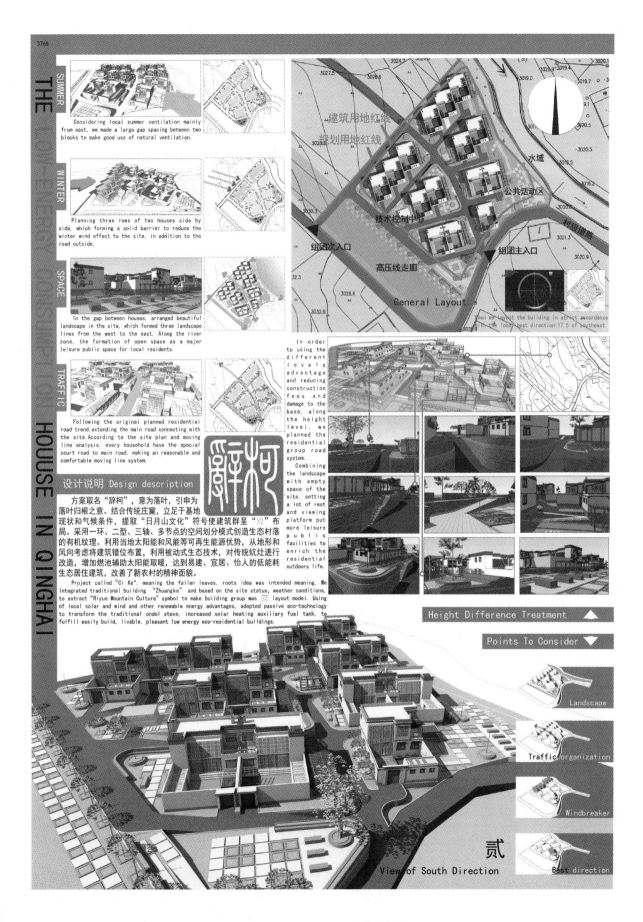

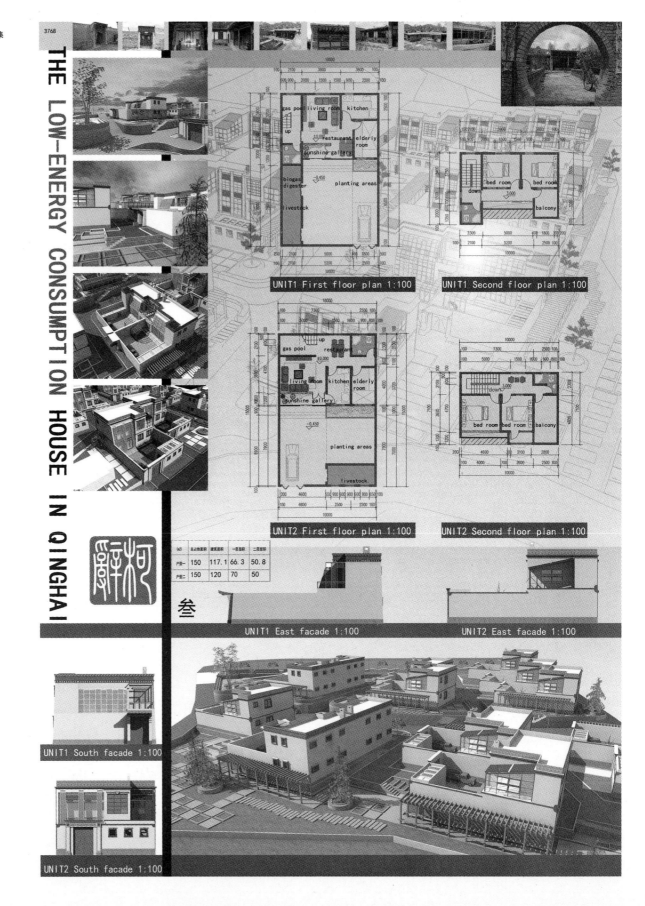

THE LOW-ENERGY CONSUMPTION HOUSE IN QINGHAI

The Concept of Ecological

Plan analysis

- Waste and polution separation
- Active and inactive space separation
- Inner space relationship
- Moving line

The change of build

- Building and courtyard
- Frame to the mass
- Enrich mass and courtyard
- Add the cultural element to building and courtyard door's head
- Integrate the thermal insulation and arounding's landscape into building

Presentations of technical

- A 冬季防风带 / Winter windbreaks
- B 太阳能墙 / Solar wall
- C 阳光间 / Solar house
- D 通风百叶 / Shutter ventilation
- E 屋顶天窗 / Sky window in roof
- F 倒置屋面 / Inverted Roof
- G 太阳能集热器 / Solar collectors
- H 地板辐射采暖 / Floor panel heating
- I 再生墙 / Regeneration wall
- J 暖房 / Green house
- K 屋面雨水收集 / Roof rainwater collection
- L 下凹式绿地收集池 / Sunken lawn collection pool
- M 燃池 / Fire Pit
- N 沼气池 / Digesters
- O 高架灶 / High stoves
- P 夏季通风道 / Ventilation entrance

Different angle views

B.I.R.D V.I.E.W P.O.I.N.T.S.

View of street

View of village house

View of river blank

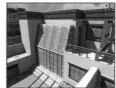
View of unit's court

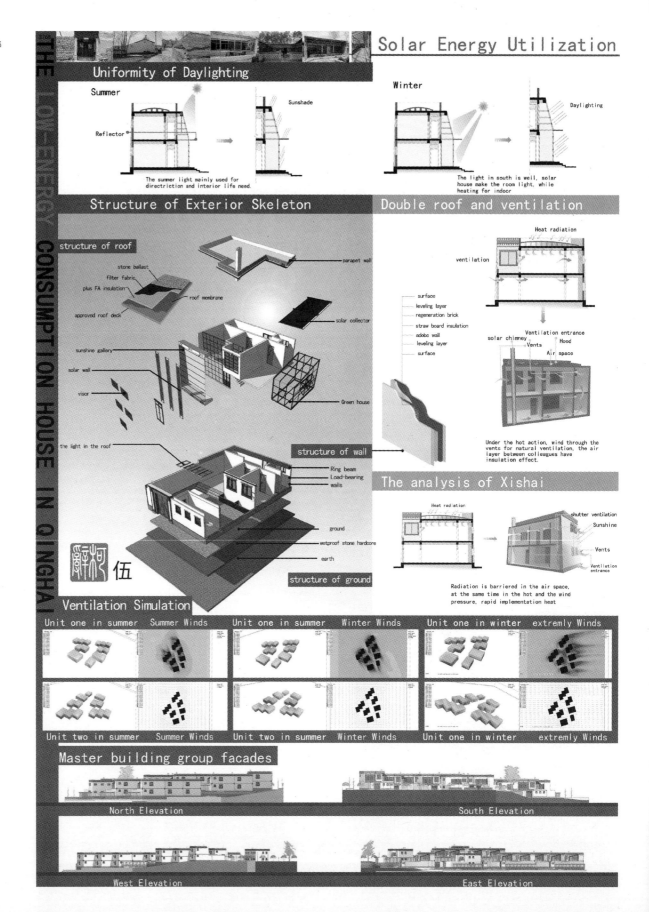

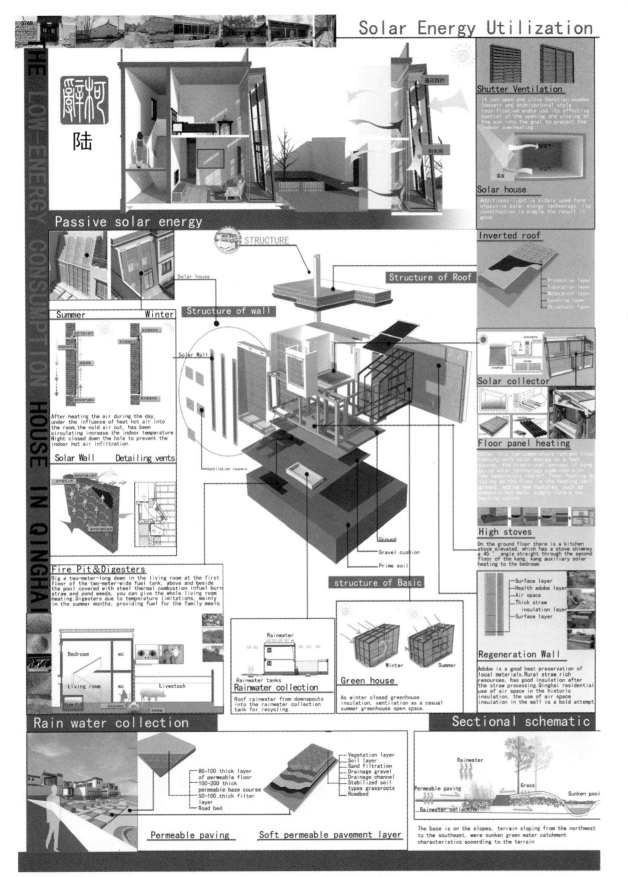

综合奖·优秀奖
General Prize Awarded · Honorable Mention Prize

注　册　号：4159
项目名称：光回故里（青海）
　　　　　Sunning House (Qinghai)
作　　者：张　宁、冯伟杰、魏　娜、
　　　　　王　云
参赛单位：大连理工大学

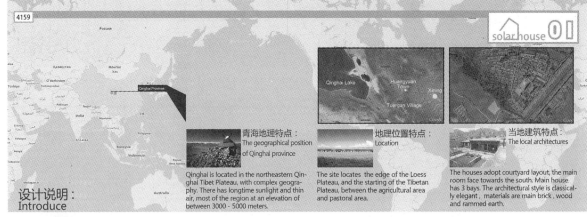

设计说明：
Introduce

本案以"光回故里"为题，描绘了对新农村住宅所寄予的美好愿景：
一是在有限条件下，对当地无限光能资源得以最大化利用，实现低能耗；二是充分尊重当地风土人情，为世代漂泊的农牧民建造足以安稳久居的家。
综合考虑气候特征，地域特性，理性思考：
1. 以太阳能与建筑一体化设计为基础。2. 规划过程中为减少土方量，节约土地，对场地作台地式处理。
3. 对关键性构造节点进行创新设计，如对新式夯土墙的构造设计等，体现节材目标。

Making Light At Home
The project, titled by making light at home, has depicted the image of rural housing, showing its double meanings under the paper. One gose to the goal of savimg energy by maximizing the use of local resources. The other goes to the hope to bring the herdsman back into a kind of safe life by buiding the comfortable shelters for them. Both considering the climate feature and the local characteristics, the design of the project is moved forward logically. 3 key techniques focus most among the all-1. Integrate the solar component with the housing buiding. 2. The planning of the site, being as terrance style saves the earth amount then to save the land. 3. Completed the structure node of some part such as the new earth wall.

光回故里
Suning Home
青海低能耗住房项目

Questions & Solutions:

Passive Solar Energy Utilization:
- The best orientation of the site: the Southeast 15°
- Heating by the passive solar house

Active Solar Energy Utilization:
- Solar energy hot water floor radiant heating system

Respect the Local Traditional:
- Using local materials: rammed earth → Houses adopt courtyard
- Traditional architecture symbol → Buildings have 3 bays

光回故里
Suning Home
青海低能耗住房项目

Plan Analysis:

Shadow Analysis
Vernal Equinox | June Solstice | Autumnal Equinox | Winter Solstice

Wind Analysis
Winter 1.500 | Winter 4.800
Summer 1.500 | Summer 4.800

Climate Analysis:

Solar Radiation
Annual incident solar radiation at 175.0°
Radiant quantity | Radiant hours

Wind Speed
Wind Frequency | Average relative humidity
Wind speed

Temperature & Humidity
Average relative humidity | Psychrometric chart
Temperature | Humidity

Rainfall
Average rainfall

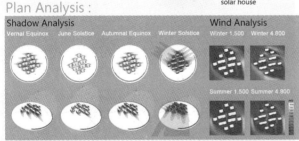
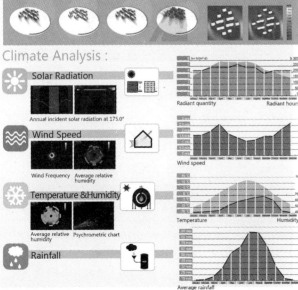

Site Plan 1:500

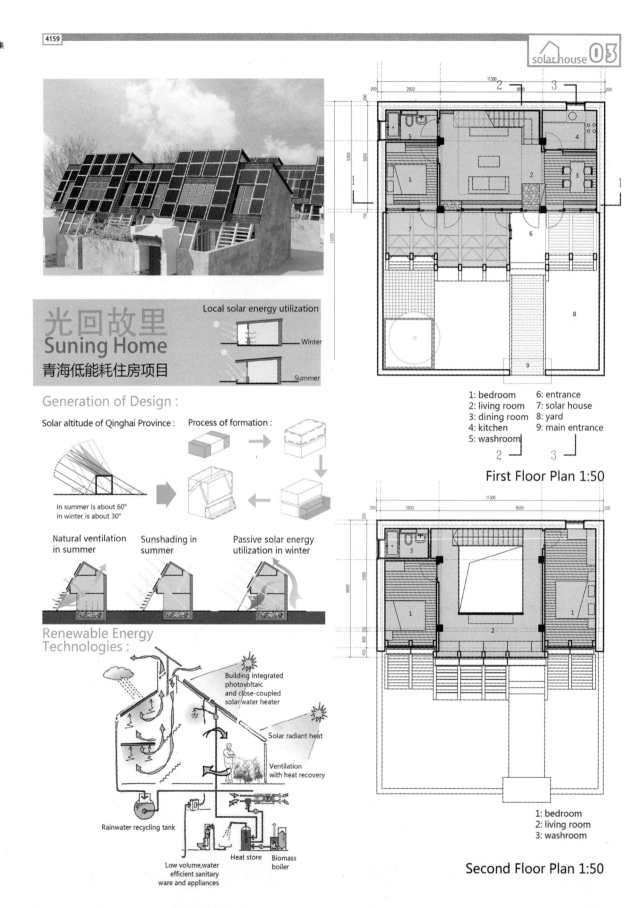

Low Temperature Hot Water Floor Radiant Heating:

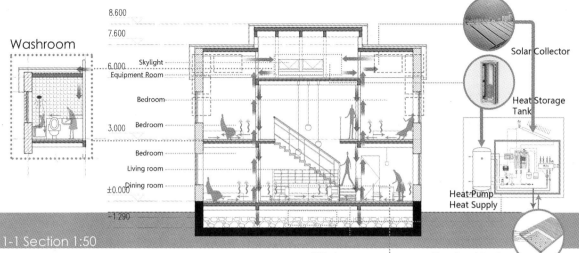

1-1 Section 1:50

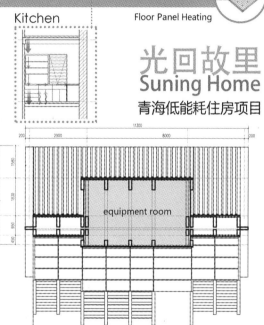

光回故里
Suning Home
青海低能耗住房项目

6.00m Plan 1:50

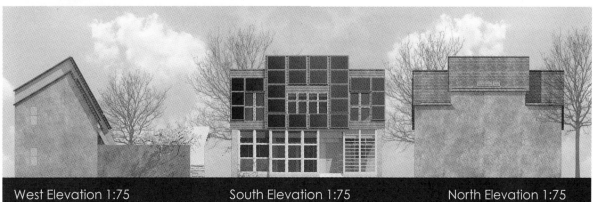

West Elevation 1:75   South Elevation 1:75   North Elevation 1:75

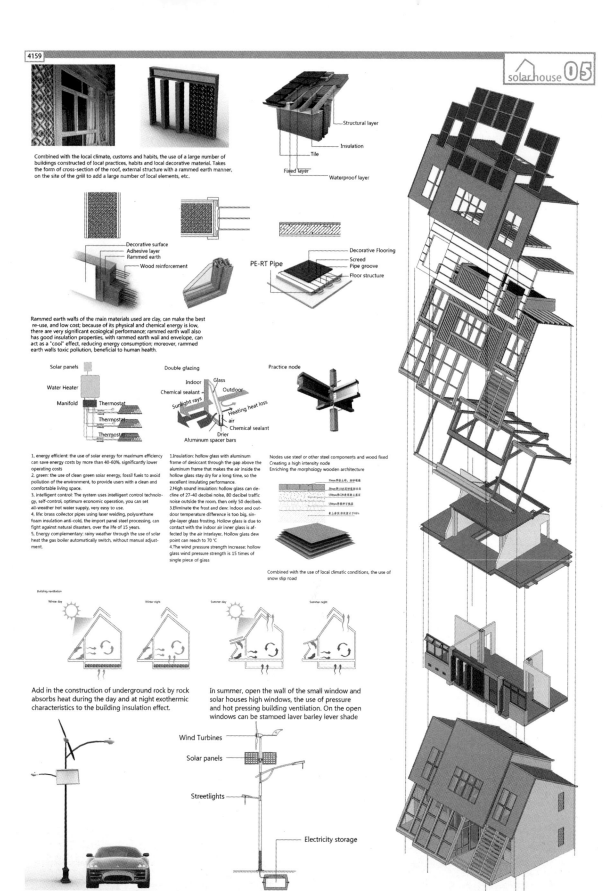

Combined with the local climate, customs and habits, the use of a large number of buildings constructed of local practices, habits and local decorative material. Takes the form of cross-section of the roof, external structure with a rammed earth manner, on the site of the grill to add a large number of local elements, etc.

Rammed earth walls of the main materials used are clay, can make the best re-use, and low cost; because of its physical and chemical energy is low, there are very significant ecological performance; rammed earth wall also has good insulation properties, with rammed earth wall and envelope, can act as a "cool" effect, reducing energy consumption; moreover, rammed earth walls toxic pollution, beneficial to human health.

1, energy efficient: the use of solar energy for maximum efficiency can save energy costs by more than 40-60%, significantly lower operating costs
2, green: the use of clean green solar energy, fossil fuels to avoid pollution of the environment, to provide users with a clean and comfortable living space.
3, intelligent control: The system uses intelligent control technology, self-control, optimum economic operation, you can set all-weather hot water supply, very easy to use.
4, life: brass collector pipes using laser welding, polyurethane foam insulation anti-cold, the import panel steel processing, can fight against natural disasters, over the life of 15 years.
5, Energy complementary: rainy weather through the use of solar heat the gas boiler automatically switch, without manual adjustment.

1.Insulation: hollow glass with aluminum frame of desiccant through the gap above the aluminum frame that makes the air inside the hollow glass stay dry for a long time, so the excellent insulating performance.
2.High sound insulation: hollow glass can decline of 27-40 decibel noise, 80 decibel traffic noise outside the room, then only 50 decibels.
3.Eliminate the frost and dew: indoor and outdoor temperature difference is too big, single-layer glass frosting. Hollow glass is due to contact with the indoor air inner glass is affected by the air interlayer, Hollow glass dew point can reach to 70 °C
4.The wind pressure strength increase: hollow glass wind pressure strength is 15 times of single piece of glass

Nodes use steel or other steel components and wood fixed
Creating a high intensity node
Enriching the morphology wooden architecture

Combined with the use of local climatic conditions, the use of snow slip road

Add in the construction of underground rock by rock absorbs heat during the day and at night exothermic characteristics to the building insulation effect.

In summer, open the wall of the small window and solar houses high windows, the use of pressure and hot pressing building ventilation. On the open windows can be stamped laver barley lever shade

光回故里
Suning Home
青海低能耗住房项目

	Function	Area		
1st floor	Living room	24.6		
	Bed room	9.2		
	Kitchen	5.8		
	Bathroom	4.5		
	Dining Room	7.8		
	Sunshine house	13.4	65.3	
2nd floor	Bedroom 1	9.6		
	Bedroom 2	13.9		
	Corridor	17.7		
	Bathroom	4.3	45.5	110.8
Garden			54.7	46.8
Total				157.6

Total Site area	9834
Floor area ratio	0.169

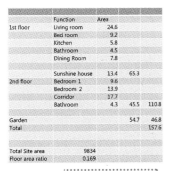
Green Façade

Traditional Symbol

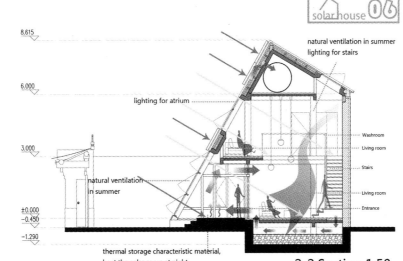
2-2 Section 1:50

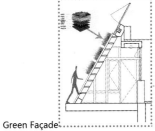
3-3 Section 1:50

Index of Building Heat Loss

以采暖季为研究对象进行估算

一、供应每日用水所需的太阳能集热器面积

1. 家庭日均用水量计算

人均日用水量为 70L　　家庭平均人口数:5

= Q_w: 家庭日均用水量人均日用量为350L

2. 太阳能集热效率

JT: 地表热器采光面上的年平均日太阳辐照量
f: 太阳保证率
ηcd: 集热器的年平均集热效率
ηL: 贮水箱和管路的热损失率

= 太阳能最后可利用的能量

3. 供应每日用水所需的太阳能集热器面积计算

$A_c = \dfrac{Q_w C_w (t_{end}-t_i) f}{J_T \eta_{cd}(1-\eta_L)}$ 经计算 = 2.2m²

二、供应每日采暖系统所需太阳能集热器面积

1. 建筑物耗热量

S: 建筑面积　　q: 采暖热指标

= Q_n: 建筑物耗热量

2. 太阳能集热效率

JT: 地表热器采光面上的年平均日太阳辐照量
f: 太阳保证率
ηcd: 集热器的年平均集热效率
ηL: 贮水箱和管路的热损失率

= 太阳能最后可利用的能量

3. 供应每日用水所需的太阳能集热器面积计算

$A_c = Q_n\, f/JT\ \eta cd$　经计算　$A_c = 27.9 m^2$

最终计算出所需的太阳能集热器面积 $A_c = 30.1 m^2$

综合奖·优秀奖
General Prize Awarded ·
Honorable Mention Prize

注 册 号：4183
项目名称：日光宝盒（青海）
　　　　　Sunshine in the Box (Qinghai)
作　　者：黄　杰、吴鑫澜、韩　凯、
　　　　　陈　诗、张嫩江、宋　祥
参赛单位：西北工业大学、西安建筑科技
　　　　　大学

日光宝盒 SUNSHINE IN THE BOX

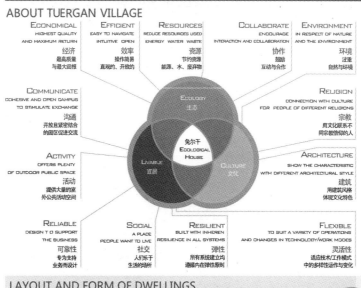

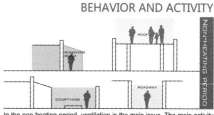

In the non-heating period, ventilation is the main issue. The main activity space in the village is on the outside. The activity space on roof platform is relatively open, and has good ventilation, lighting and landscape environment.

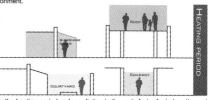

In the heating period, solar radiation is the main factor for indoor thermal environment. During the day, the roof and the solar house with good orientation of the building absorb solar radiation. While the courtyard and street is easy to form shadow space. At night, the solar house changes to buffer layer to keep warm.

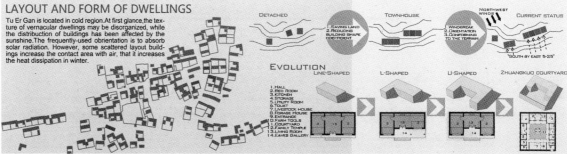

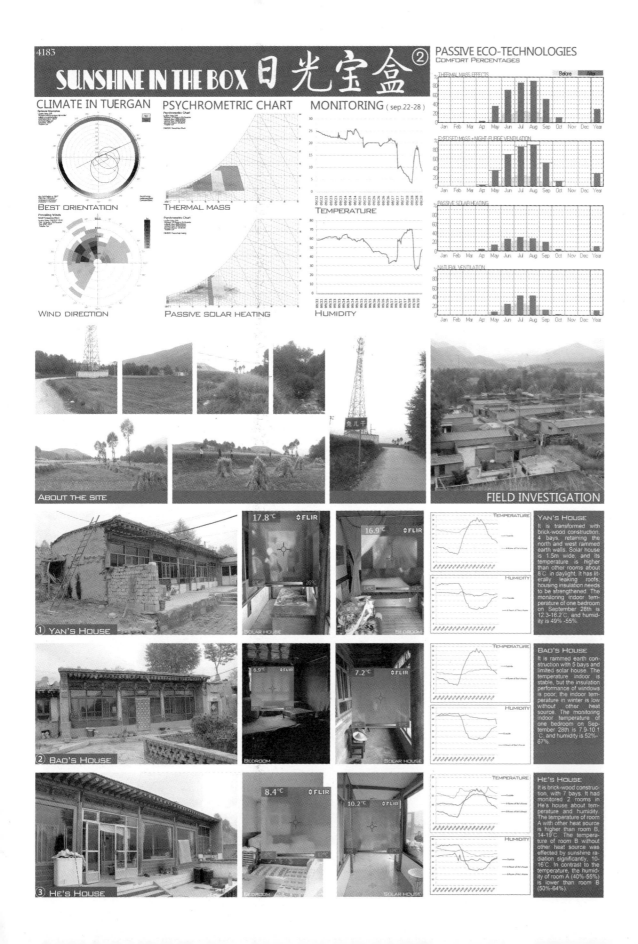

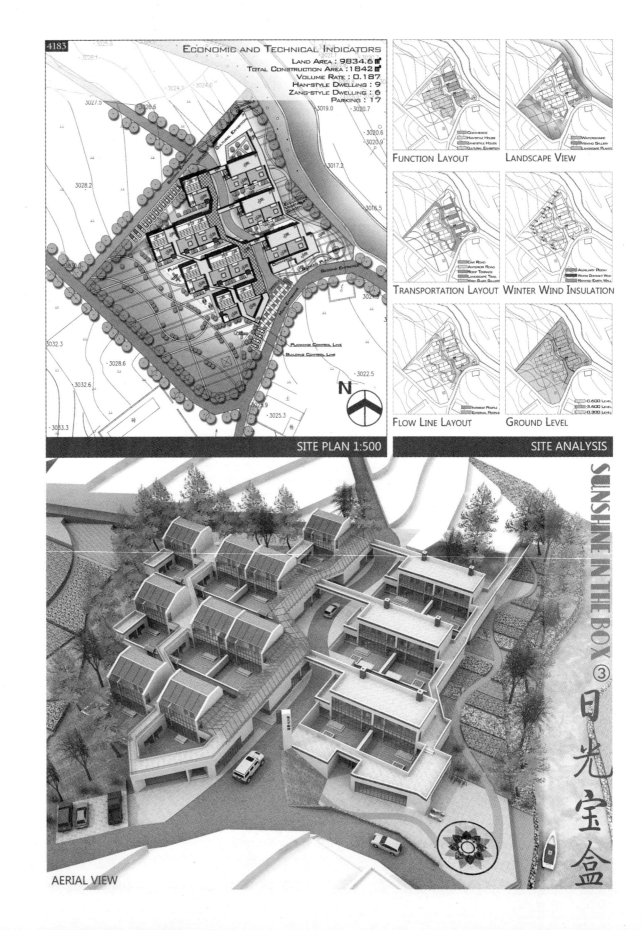

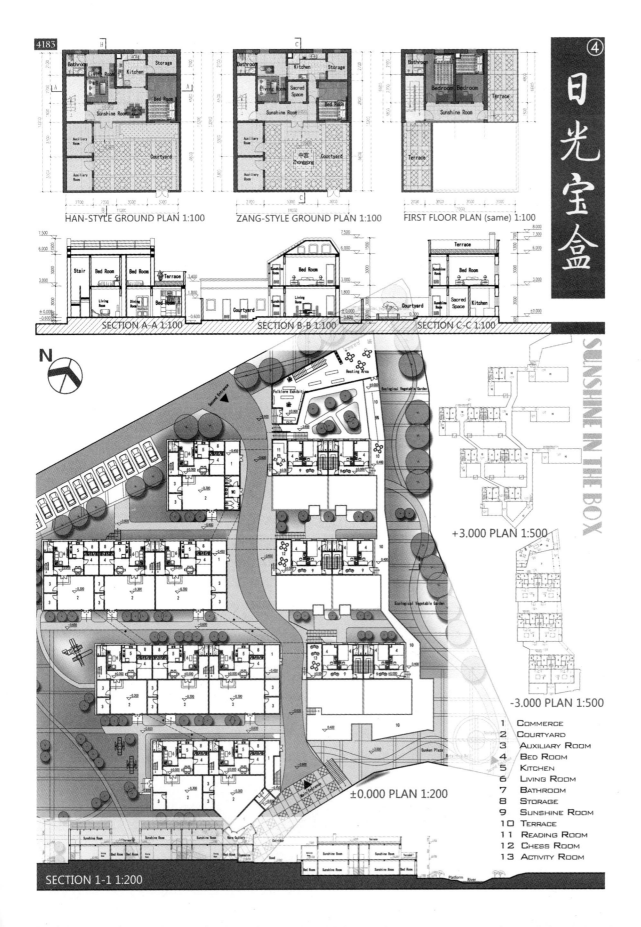

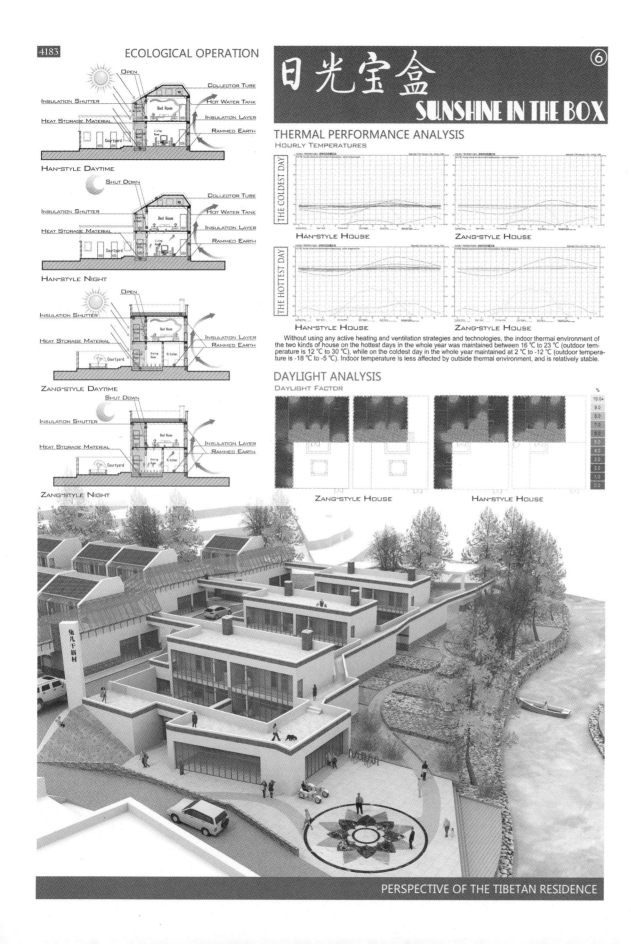

SUNSHINE IN THE BOX
日光宝盒

ECOLOGICAL OPERATION

THERMAL PERFORMANCE ANALYSIS
Hourly Temperatures

Without using any active heating and ventilation strategies and technologies, the indoor thermal environment of the two kinds of house on the hottest days in the whole year was maintained between 16 ℃ to 23 ℃ (outdoor temperature is 12 ℃ to 30 ℃), while on the coldest day in the whole year maintained at 2 ℃ to -12 ℃ (outdoor temperature is -18 ℃ to -5 ℃). Indoor temperature is less affected by outside thermal environment, and is relatively stable.

DAYLIGHT ANALYSIS
Daylight Factor

PERSPECTIVE OF THE TIBETAN RESIDENCE

综合奖·优秀奖
General Prize Awarded·
Honorable Mention Prize

注 册 号：4347
项目名称：土筑春意（青海）
　　　　　Green Residence Built by
　　　　　Local Mud
作　　者：胡琪蔓、向奕妍、张　璐、
　　　　　冯　昱、杨巧霞
参赛单位：重庆大学

土筑春意
2015台达杯太阳能建筑设计大赛

CHAPTER 1　PRE-PROGRAM

1.1 Background analysis

1.1.1 GEOGRAPHY AND CLIMATE ANALYSIS

XINING

RIYUE ZANG XIANG

THE ANNUAL AVEIAGE SUNSHINE HOURS　2674H
THE ANNUAL AVERAGE TEMPERATURE　10.3 C°
THE ANNUAL HUMIDITY　49%
THE ANNUAL WIND SPEED　0.85M/S

Perfect sunshine: conductive to the use of solar energy.
Good temperature: the annual average temperature is good, but there have hot summer and cold winter in QIN HAI province.
Bad humidity: low air humidity, easy to get dry.
Nice wind: moderate speed, pleasant breeze.

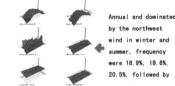
Annual and dominated by the northwest wind in winter and summer, frequency were 18.9%, 19.6%, 20.5%, followed by

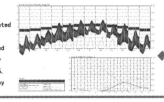
The green line represents the thermal comfort zones, October 4 - more comfortable, in other uncomfortable, remains to be

THE DESIGN DESCRIPTION
THE MOST DISTINCTIVE FEATURE OF THE BUILDING IS TO LOCALLY SOURCED AND GREEN-ENERGY CONSERVATION. THE ARCHITECTURAL FROM HAS LOCAL FLAVOR BECAUSE IT INHERIT FROM THE LOCAL TRADITIONAL HOUSES (ZHUANG KUO) AND COMBINE WITH LOCAL BUILDING MATERIALS (SUCH AS COBBLESTONE). TWO KINDS OF HOUSE TYPE ADAPTED TO THE DIFFERENT TYPES OF FAMILIES. THE HOUSE TYPE WHICH LIKES A KEY IS SUITED TO LIVING FOR THE FAMILY OF THREE GENERATIONS. AND THE ROW HOUSE IS SUITED TO LIVING FOR YONG COUPLES.

设计说明
建筑最大的特色在于取材当地，绿色节能。平面脱胎于当地传统庄廓，结合当地的鹅卵石等材料，形成古朴乡土的特色。两种户型分别适应于不同结构的家庭，钥匙头户型较大，适应于三代同堂的家庭居住；联排户型紧凑节地，适应于年轻夫妇居住。

1.2 Current analysis

1.2.1 CITY IMAGE ANALYSIS

1. Rabbit's dry village in qinghai province in the Tibetan plateau, deep in the inland, away from the ocean, the continental climate belongs to the plateau.
2. Huangyuan county deep in the inland, is a continental climate, the illumination time is long, strong solar radiation, the temperature difference big, windy spring, summer is cool, as the main direction, northwest dry winter, frost-free period is short, hail, drought frequently.
3. The sun township, the whole landscape is given priority to with original alpine valleys and mountain area, domestic water resource is very rich, has a prestigious Tibetan Buddhism temple — domhke temple.

1.2.2 LOCAL ARCHITECTURAL STYLE ANALYSIS

1. Wall outside wall of build by laying bricks or stones into trapezoidal, strong thick, general will be higher than the internal building about half a meter to withstand cold.
2. Exterior wall only open a door, with strong defensive.
3. "Zhuang kuo" is influenced by central plains culture of the han nationality, so the yard in the form of courtyard (SIHEYUAN).
4. Biggest due north direction room, to the highest position elders live in the home, what wing is located in the left and right sides.
5. "Zhuang kuo" generally adopt overhangs flat roof form, the roof can be placed to dry goods, also can defense foreign enemy.

1.3 Behavior analysis

1.3.1 RELIGIOUS BELIEF

Tibetan Buddhism

1.3.2 PRODUCTION AND LIVING BEHAVIOR

The main crops are wheat, Highland barley, Broad beans, the potato, rape and so on. They all need to dry, so we need to design the roof terrace.

In the process of agricultural production need to use the harvester farm tools, such as grazing activities at the same time, so consider the design of livestock.

CHAPTER 2 PLANNING DESIGN

2.1 Concept analysis

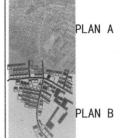

PLAN A

PLAN B

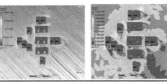

It can be seen from the simulation results:
In project 1, the local wind speed in the west of community exceeds 1.5m/s, and there are no wind zones behind the buildings. Eddy current field is relatively serious.
In project 2, the local wind speed in the northwest of community exceeds 1.5m/s. As the predominant wind direction is westerly, it is recommended to plant tall trees in the northwest edge for the resistance of winter wind. It can be seen from the wind field distribution that in project 2, the no wind zones in the lee side of buildings are less than that in project 1, there is almost without no wind zone, also without eddy current, and the wind speed distribution is relatively uniform, which is advantage for the diffusion of pollutants such as livestock manure.

PLAN B

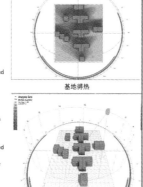

ECOTEC SIMULATION

基地得热

南立面得热

2.2 Master plan analysis

Economic and technical norms:
1. Site area: 4999.5㎡
2. Coverd area: 1555.2㎡
3. Overall floorage: 3160㎡
4. Building density: 31.1%
5. Floor area ratio: 0.632

MASTER PLAN 1:1000

 阳光间与太阳能分布

 生态绿化层级理念

 道路分析

 生态绿化点线面

2.3 Aeviar view

土筑春意
2015台达杯太阳能建筑设计大赛

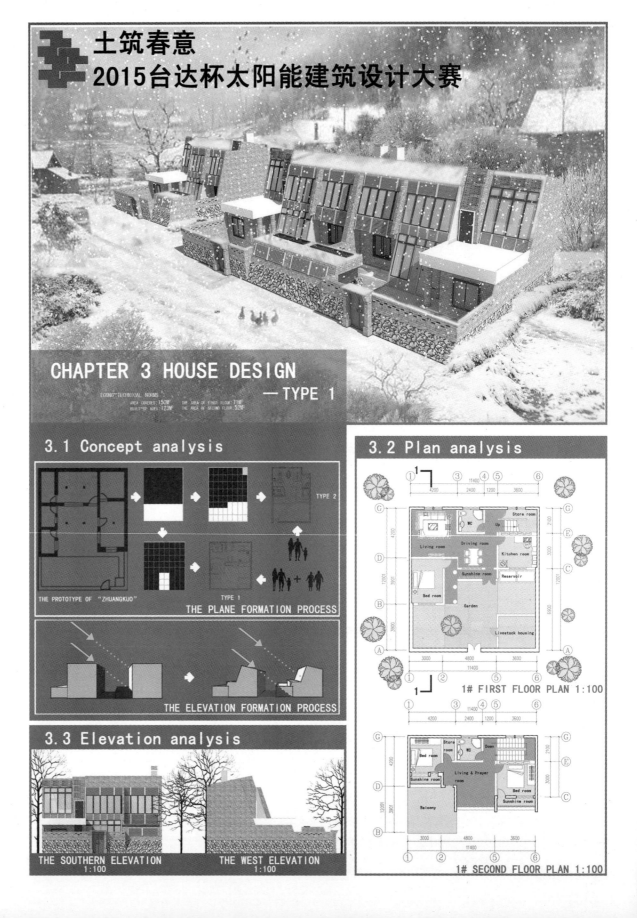

土筑春意
2015台达杯太阳能建筑设计大赛

CHAPTER 4 HOUSE DESIGN —TYPE 2

ECONO-TECHNICAL NORMS
AREA COVERED: 150M²   THE AREA OF FIRST FLOOR: 68M²
BUILT-UP AREA: 115M²   THE AREA OF SECOND FLOOR: 47M²

4.1 Plan analysis

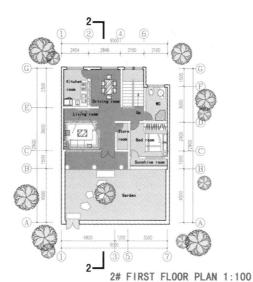

2# FIRST FLOOR PLAN 1:100

4.2 Elevation & section analysis

THE SOUTHERN ELEVATION 1:100

THE EAST ELEVATION 1:100

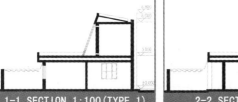
1-1 SECTION 1:100 (TYPE 1)   2-2 SECTION 1:100

2# SECOND FLOOR PLAN 1:100

1# & 2# ROOF PLAN 1:1000

4.3 Aerial view

土筑春意
2015台达杯太阳能建筑设计大赛

AEVIAL VIEW

Square perspective

The terrace housing local perspective

Roof terrace

CHAPTER 5 BUILDING PHYSICAL ENVIRONMENT ANALYSIS

5.2 Ventilatte analysis

Winter Day | Summer Day
Winter Night | Summer Night

winter
We append a solarium near the bedroom, and use heat-trapping wall with holes up and down to keep the solarium and the bedroom separate. In the daytime, the heat-trapping wall translate solar energy into heating and store it. At night, the heat-trapping wall use heating to improve indoor temperature.

summer
In summer, use the roof with planting lawn to prevent heat from coming into rooms. The flat roof can keep the sun out of rooms. The controllable shutters can accelerate air currents of solarium and take away heat.

5.1 Sunlight analysis

The summer solstice at 12 noon
The winter solstice at 12 noon
The summer solstice at 12 noon
The winter solstice at 12 noon

技术剖视图 Technical sections

- 绿化屋顶 green roof
- ⑦阳光间 The sun heat storage wall (attached)
- ⑥可控百叶窗(内附玻璃) Controlled blinds (inside the glass)
- ①太阳能拔风烟囱 The solar wind pull the chimney
- ③火墙 Fire wall
- ④低温热水辐射地板 Low Tm hot water floor
- ⑤辐射玻璃 LOW-E glass
- 10鹅卵石床蓄热系统 Pebbles heat storage bed system
- ⑧屋顶和墙体雨水收集系统 Roof, walls catchment system
- ⑨牲畜间采暖 Heating of the livestock

CHAPTER 6 TECHNIQUE

6.1 Energy saving technology overview

该建筑所处位置的气候为:昼夜温差大,夏季凉爽,冬季干燥寒冷。因此,该建筑的热环境应以保温为主。隔热为辅。结合当地特色,共设计了10多种措施。

The climate of the location is: big temperature gap between day and night, summer is cool, dry and cold in winter. Therefore, the building thermal environment should be given priority to with heat preservation. Heat insulation is complementary. Combined with local characteristics, a design of more than 10 kinds of measures.

时间	措施
夏季 Summer	绿化屋顶 Green roof
	可控百叶窗 Controlled blinds
	Low-E 玻璃 Low-E Glass
冬季 Winter	绿化屋顶 Green roof
	可控百叶窗 Controlled blinds
	阳光间 The sun heat storage wall
	火墙 Fire Wall
	低温热水辐射地板 Low Tm hot water floor
	鹅卵石床蓄热系统 Pebbles heat storage bed system
	牲畜间采暖 Heating of the livestock
四季 Four seasons	屋顶和墙体雨水收集系统 Roof,walls catchment system
	太阳能光伏电板 Solar photovoltaic panels
	太阳能拔风烟囱 The solar wind pull the chimney

Passive solar
Energy other
active solar

土筑春意
2015台达杯太阳能建筑设计大赛

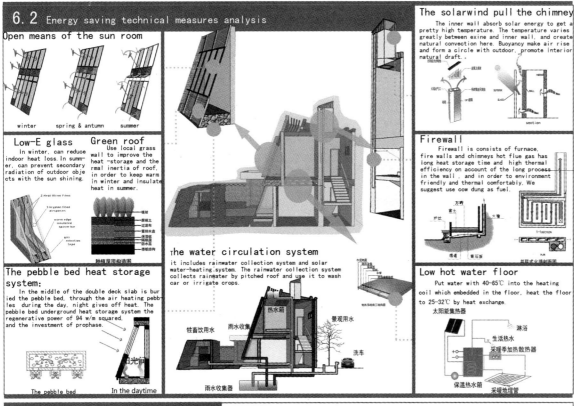

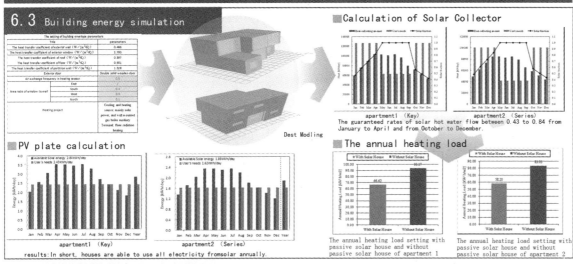

AEVIAR VIEW — The south of buildings

综合奖·优秀奖
General Prize Awarded · Honorable Mention Prize

注 册 号：4372
项目名称：一间阳光（青海）
　　　　　Rooms of Sunshine (Qinghai)
作　　者：陈 莉、柴克非、李 垚、
　　　　　薛 凯
参赛单位：重庆大学

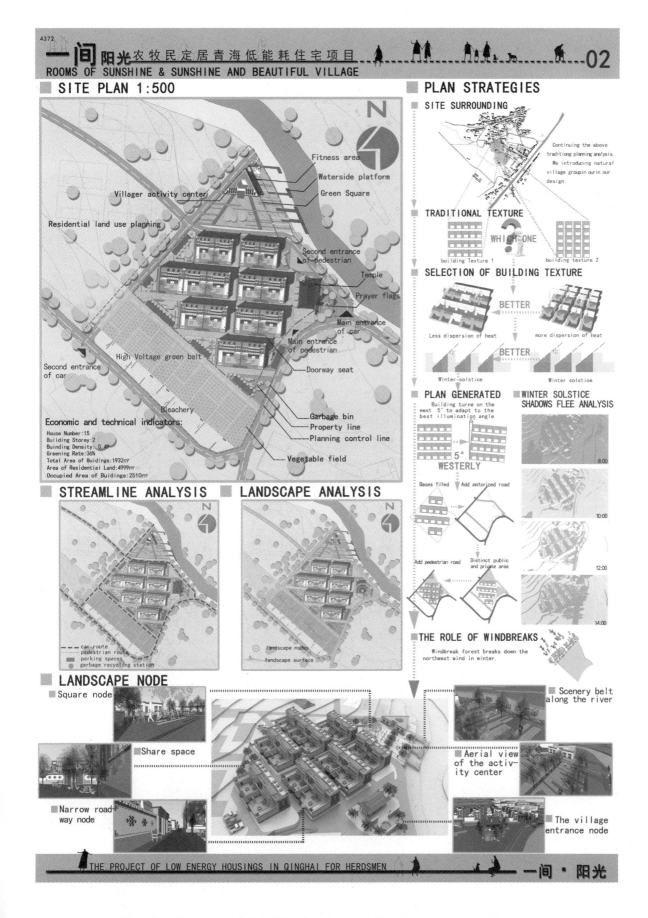

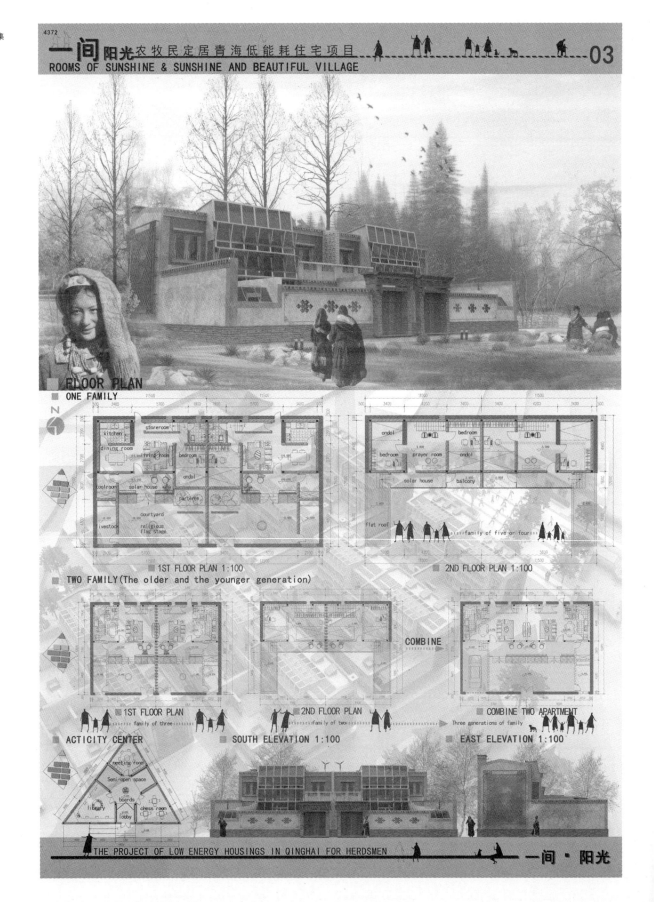

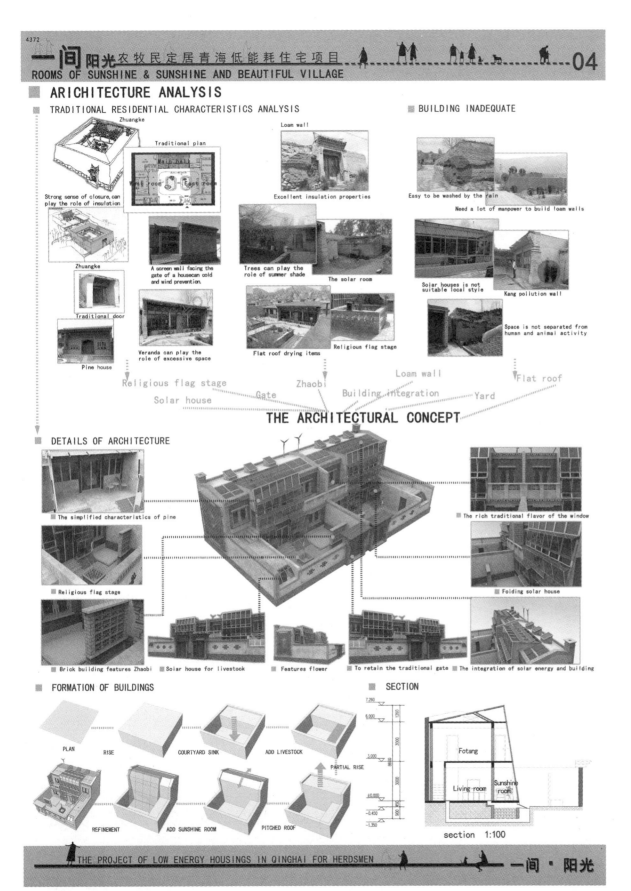

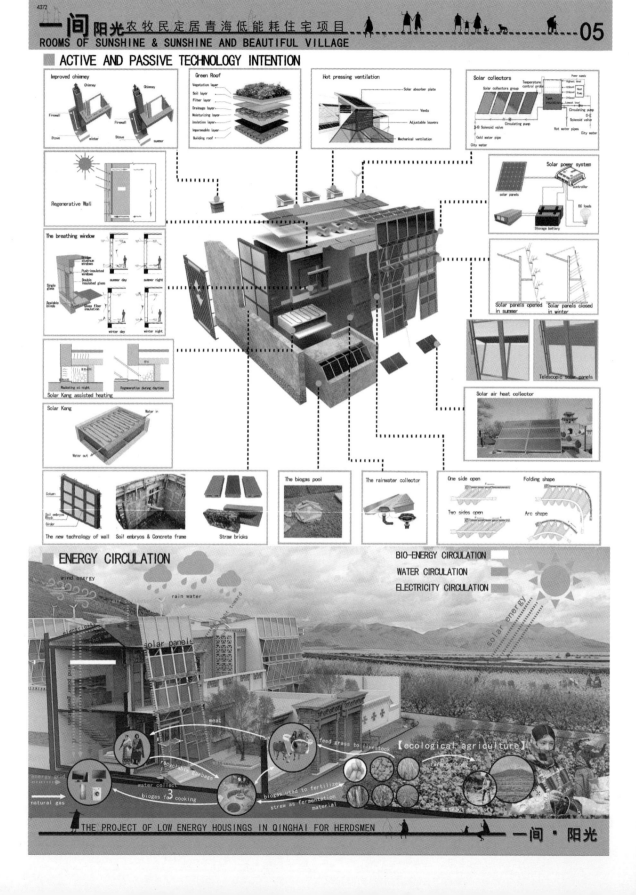

ROOMS OF SUNSHINE & SUNSHINE AND BEAUTIFUL VILLAGE
一间阳光 农牧民定居青海低能耗住宅项目

UNIVERSALITY OF SOLAR

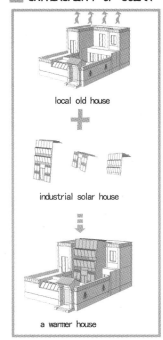

local old house
+
industrial solar house
↓
a warmer house

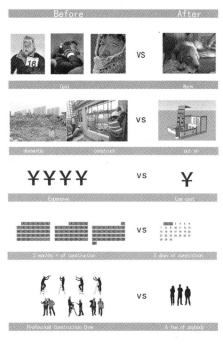

Before vs After
Cold vs Warm
dismantle · construct · built on
¥¥¥¥ vs ¥
Expensive vs Low cost
3 months of construction vs 2 days of construction
Professional Construction Crew vs A few of anybody

SCHEMATIC CHANGES OF SOLAR

OPEN

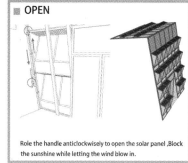

Role the handle anticlockwisely to open the solar panel, Block the sunshine while letting the wind blow in.

CLOSE

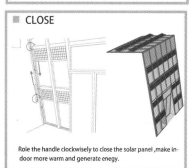

Role the handle clockwisely to close the solar panel, make indoor more warm and generate enegy.

ANALYSIS OF THE SUMMER AND WINTER

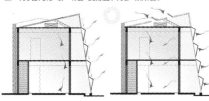

Summer day, open the solar panels up, then the sunshine will be protect outside of the sunshine room, and will creat more electric, so the temperature indoors will be lower than normal buildings.

Summer night, hole open, indoor high pressuer hot air is brought out, reduce the room temperature.

Winter day, the sun heating the sunshine room, then heat the storage wall. The heat will transfer to indoors, and the cold air will become warm.

Winter night heat storage wall during the day will be absorbed by the heat rejection, thermal insulation layer to prevent indor heat flow and prevent cold air into the outdoor.

ROOF HEATING LAYER SCHEMATIC

Air heating circulation system

Winter sun during the day through the top of the movable plate into the roof space, illuminated and heated air collector insulation, and through the air during the double wall heat into interior layer, and then passed through a gradual cooling of indoor air underground rock bed, heating cycle.

DATA ANALYSIS

plant configuration analysis
Solar radiation is between 0.4-5.4MJ/m²·d in the whole space. Combining the characteristics of local plants, planting deciduous trees mainly.

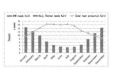

effectiveness analysis of solar panels
Heat production is insufficient to meet demand, the need auxiliary heating in winter; heat production than demand in summer.

the wind environment of first floor
The maximum wind speed is 0.4m/s. Wind through the room in the state of all windows opend. Kitchen and bathroom location also ensures the wind does not pass in the bedroom.

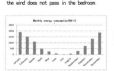

monthly energy consumption
The colder the outdoor temperature, the higher energy consumption.

the wind environment of second floor
Maximum wind speed is 0.8m/s. Winds flow from the left bedroom to the right bedroom, form a good wind environment.

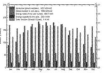

solar panels power analysis
The annual electricity demand of 680kW. Type Selection: PV panels to select seven 150W, 28V photovoltaic panels: a power of 0.9kW inverter; and the system power supply to the users about the actual 282.4kW·h, guaranteed rate of 41.5%, and the entire system power into the grid for 1192.8 kW·h, so the amount of power required is approximately 398kW·h. The average daily consumption of 3.8kW·h generated per unit installed power.

terrain and daylighting analysis
There is cover on the base, when the solar elevation angle of ~45 degrees.

SUNSHINE DEODORANT WAY BETWEEN LIVESTOCK

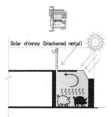

Solar chimney (blackened metal)

Solar chimney is in the air after leaving the house in It is heated, the ultimate effect of enhancing the chimney effect, Even in the absence of a windy day, when the sun shines. When shooting, rside there will be no odor.

THE PROJECT OF LOW ENERGY HOUSINGS IN QINGHAI FOR HERDSMEN — 一间·阳光

综合奖·优秀奖
General Prize Awarded · Honorable Mention Prize

注 册 号：4439
项目名称：双层聚落（青海）
　　　　　Double-deck Settlement (Qinghai)
作　　者：丁磊、陈思源、张文、张晓斐
参赛单位：天津大学、青岛绿城建筑设计有限公司

双层聚落——农牧民定居青海低能耗住房项目
Double-deck settlement

Background analysis
Location analysis

Climate analysis

Content	Count	Measures
The annual average sunshing hours	2718.6 h	Use the solar energy
The annual average solar radiation	4.67kW·h/(m²·d)	Use the solar energy
The annual average temperature	1.8℃	Thermal insulation
Monthly maximum temperature difference	37.6℃	Heat storage material
The annual average wind speed	1.73 m	Windbreaks

Traditional residence analysis

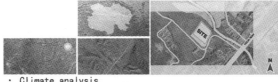

Designing description

Design concept

This design is based on the conception of "Double-deck settlement", aiming at designing the regional residential building which is of comfort livability and also low energy consumption.
"Double-Deck"
The plan originates from the inspiration of Tibetan dwellings "watchtower" and huangtu resident cave houses. To adapt to local climate, buildings are designed as folded to economize land. Taking the advantage of the terrain elevation, roads are also designed as "Double-Deck" to solve the traffic problem.
"Collective Settlement"
Folded site layout can offer a collective leisure space for the communication of local residents. In order to lower building energy consumption, improve interior environment(wind & luminous), a solar atrium which could gain solar energy is designed for each household, so that it could offer a comfortable activity space for each family.

方案以"双层聚落"为理念，设计宜居、舒适、低能耗的地域性民居建筑。
"双层"
方案以藏族民居"碉楼"和黄土高原民居窑洞为原型，适应当地的气候特点。将建筑叠合起来，节约用地，并利用地形高差，设计双层道路，解决交通问题。
"聚"
利用叠合式布局，节约用地，围合出可供人们聚集、休闲交流的空间。
嵌入"阳光中庭"，被动聚集太阳，降低建筑能耗，改善室内风光环境，提供舒适的家庭活动空间。

Space and form analysis

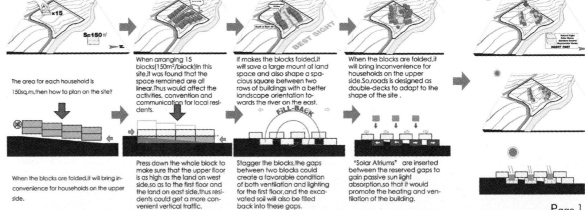

Page 1

双层聚落——农牧民定居青海低能耗住房项目
Double-deck settlement

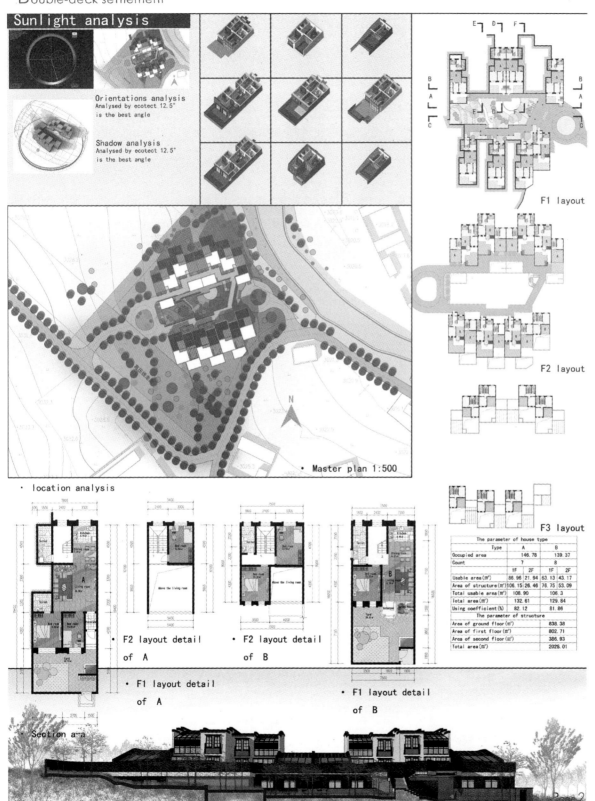

双层聚落 ——农牧民定居青海低能耗住房项目
Double-deck settlement

- Thermal analysis
- Wind simulation
- Ventilation analysis
- Section perspective

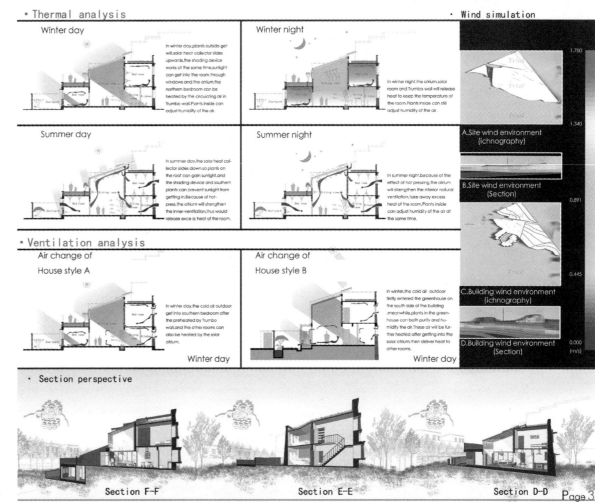

Section F-F   Section E-E   Section D-D

双层聚落 —— 农牧民定居青海低能耗住房项目
Double-deck settlement

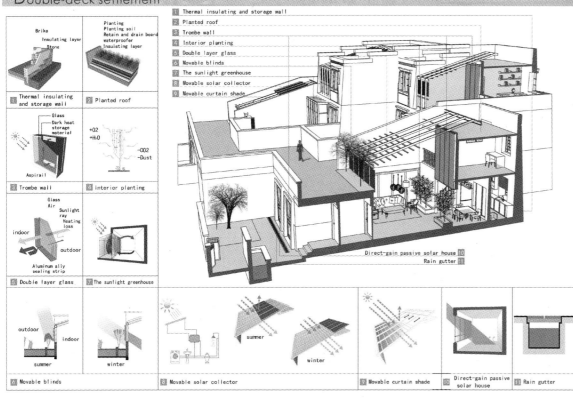

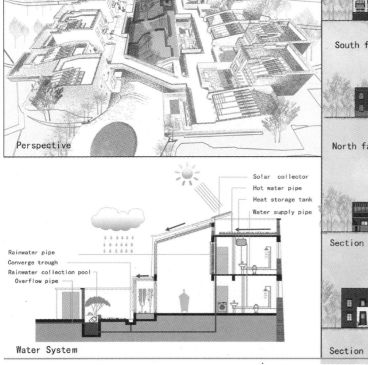

Perspective

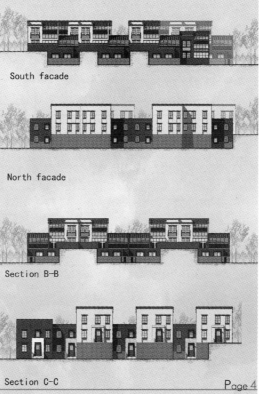

South facade

North facade

Section B-B

Section C-C

Water System

综合奖·优秀奖
General Prize Awarded·
Honorable Mention Prize

注 册 号：4466
项目名称：荒漠中的暖屋（青海）
　　　　　Warm House (Qinghai)
作　　者：孙姣姣、傅嘉言
参赛单位：浙江大学

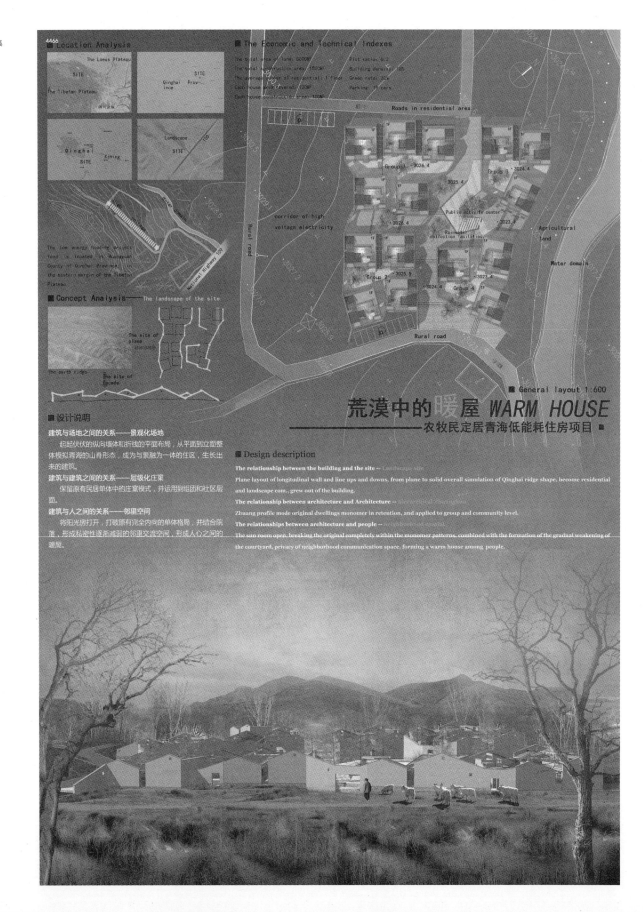

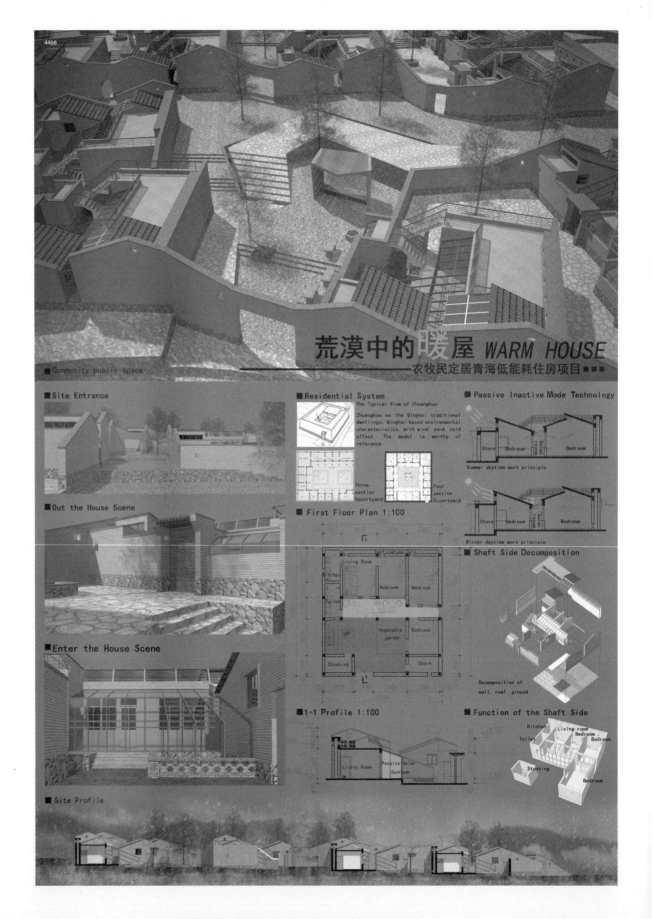

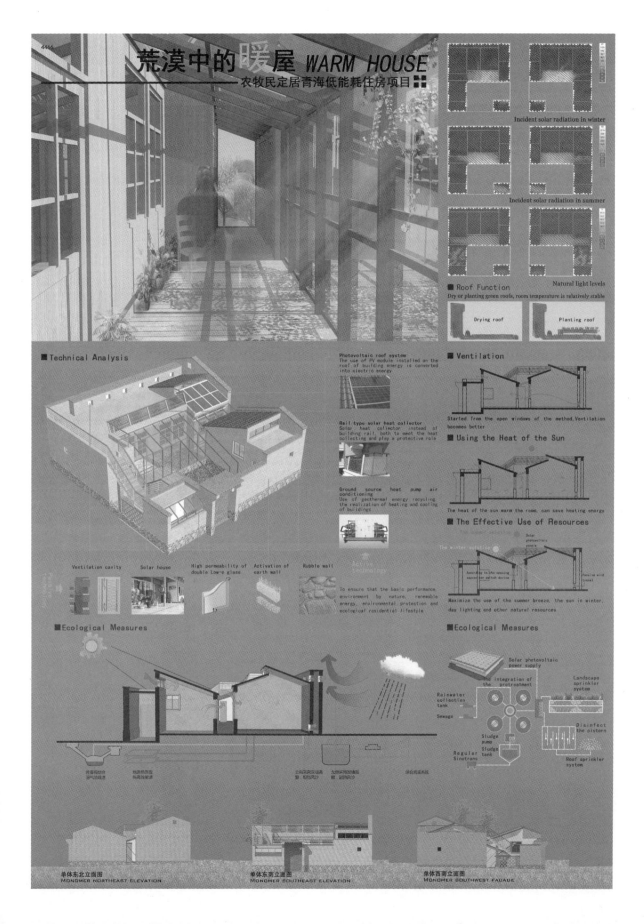

综合奖·优秀奖
General Prize Awarded · Honorable Mention Prize

注 册 号：3493
项目名称：乡村的守护与反哺（黄石）
　　　　　Metabolism System of Village (Huangshi)
作　　者：徐慧敏、郑繁硕、鞠　婧、徐睿鹏
参赛单位：山东建筑大学、Hochschule Anhalt（Bauhaus-University）

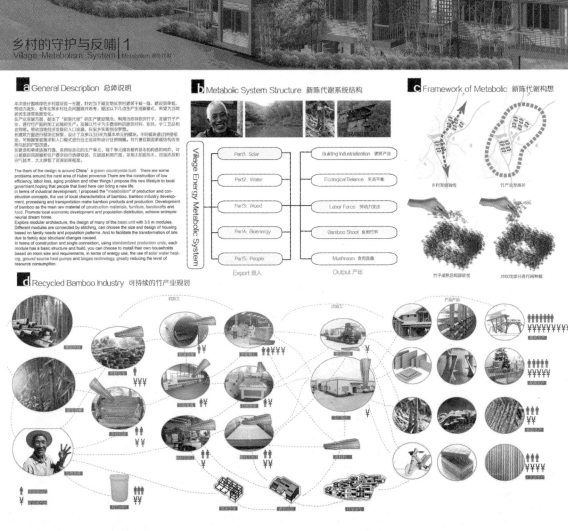

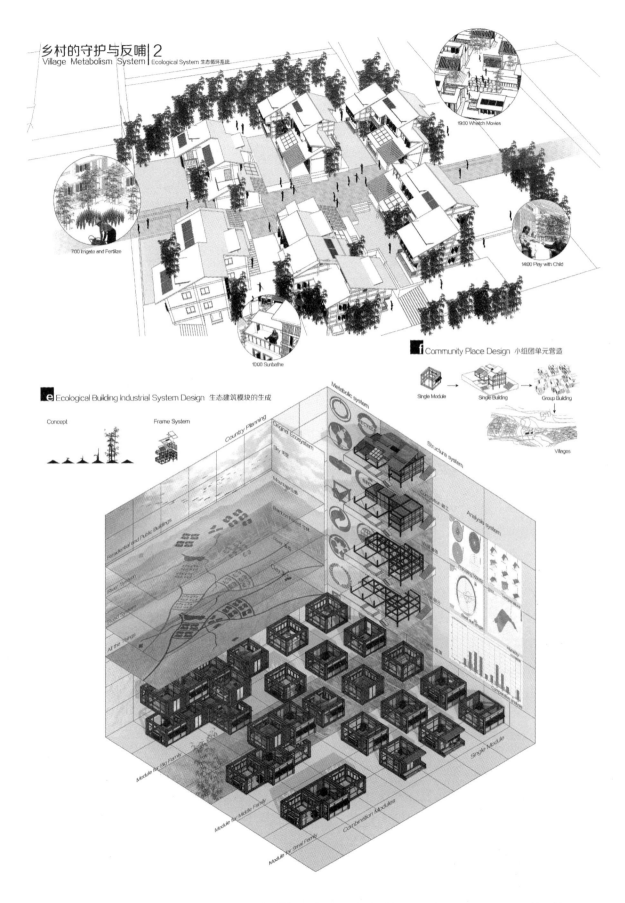

乡村的守护与反哺 | 4
Village Metabolism System | Industrial System 标准化模块

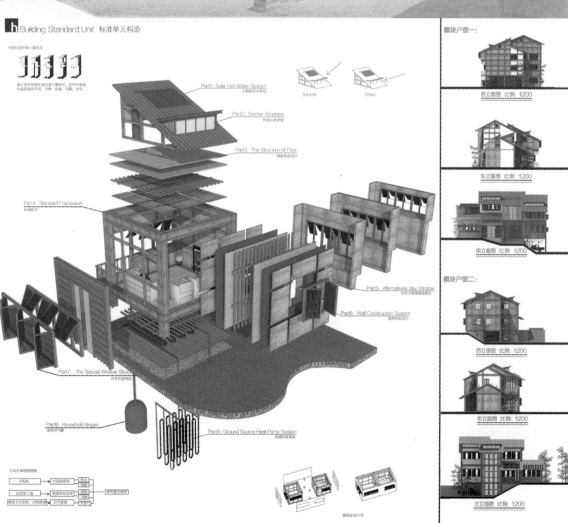

综合奖·优秀奖
General Prize Awarded·
Honorable Mention Prize

注 册 号：3534
项目名称：光井·模宅（黄石）
　　　　　Solar Patio & Module House
　　　　　(Huangshi)
作　　者：刘成威、罗杰
参赛单位：西南交通大学

SOLAR PATIO & MODULE HOUSE
CODE: 3534
Rural housing industrialization Yellowstone project

光井·模宅
农村住房产业化黄石项目

光井·模宅 — 传统平面研究与功能模块生成 Traditional Plan Research and Function Module Generation

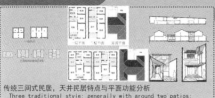

传统三间式民居，天井民居特点与平面功能分析
Three traditional style: generally with around two patios; the front patio is after the entrance; the living room (main room) in the middle of the house; after the stair, on both sides of the living room are the bedrooms; both sides of the rear patio for ancillary room; an area larger after patio to face the back alleys.

- The rooms scale is randomness, it is not conducive for industrial production.
- Room features are rigid, it will not well adapt to the modern life.

step1: 模数提取

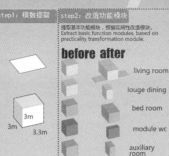

step2: 改造功能模块
提取基本功能模块，根据实用性改造模块。
Extract basic function modules, based on practicality transformation module.

before → after
- living room
- louge dining
- bed room
- module wc
- auxiliary room

step3: 组合阳光井

- living room + solar patio
- living room + louge + bed room + solar patio
- bed room + solar patio
- living room + louge + solar patio

step4: 生成3种方案

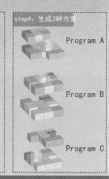

- Program A
- Program B
- Program C

光井·模宅 — 根据不同家庭需求生成方案 Generate Solutions According to Different Needs

方案A — 经商的农村家庭 Rural commercial family

The dwelling relatively owns larger land area, so the interior space for public activities is more adequate, the kind of space combined closely with the sun patio and private yard, this do good to communication activities with the commercial friend.

Front yard is more open, for communicating with people, and planting fruit trees.

rest in the dry stage

base area	167.9 m²
building area	139.1 m²
total building area	225.3 m²
building density	83%
plot ratio	1.35

Backyard for private activities

solar patio to provide heat for the interior space

First floor plan / Second floor plan

方案B — 传统的农村家庭 Traditional rural family

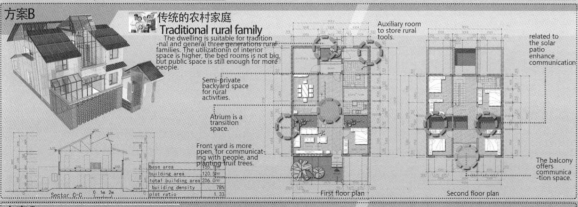

The dwelling is suitable for traditional and general three generations rural families. The utilization of interior space is higher, the bed rooms is not big, but public space is still enough for more people.

Semi-private backyard space for rural activities.

Atrium is a transition space.

Front yard is more open, for communicating with people, and planting fruit trees.

base area	154.4 m²
building area	120.5 m²
total building area	206.0 m²
building density	78%
plot ratio	1.33

Auxiliary room to store rural tools.

related to the solar patio enhance communication

The balcony offers communication space.

First floor plan / Second floor plan

方案C — 留守儿童和新兴的家庭 Stay-at-home children and seperated rural family

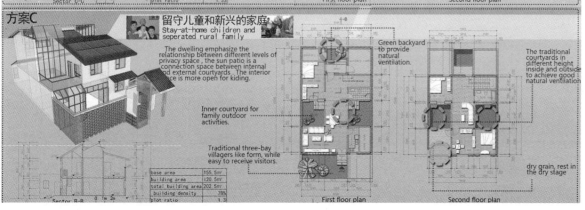

The dwelling emphasize the relationship between different levels of privacy space, the sun patio is a connection space between internal and external courtyards. The interior space is more open for kiding.

Inner courtyard for family outdoor activities.

Traditional three-bay villagers like form, while easy to receive visitors.

base area	155.5 m²
building area	120.5 m²
total building area	202.5 m²
building density	78%
plot ratio	1.3

Green backyard to provide natural ventilation.

The traditional courtyards in different height inside and outside to achieve good natural ventilation

dry grain, rest in the dry stage

First floor plan / Second floor plan

SOLAR PATIO & MODULE HOUSE
CODE: 3534
Rural housing industrialization Yellowstone project

光井·模宅
农村住房产业化黄石项目 3/6

光井·模宅 — 阳光井及构造 Solar patio construction

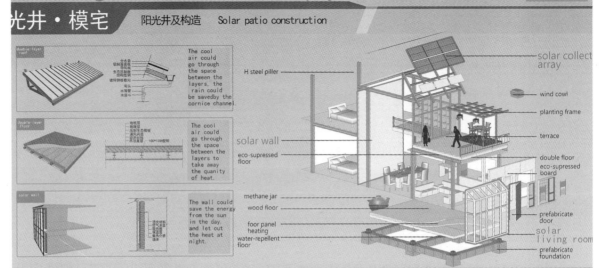

光井·模宅 — 农村住宅生态策略之原理解析 Rural ecological house analytical principle

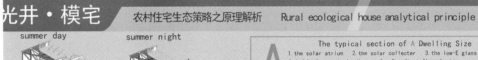

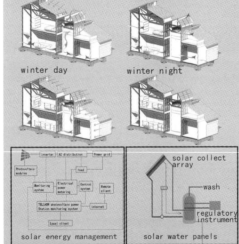

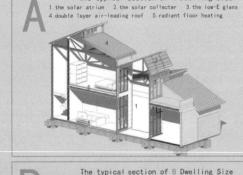

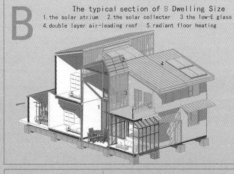

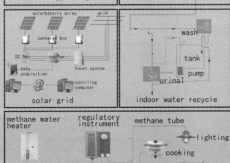

SOLAR PATIO & MODULE HOUSE
CODE: 3534
Rural housing industrialization Yellowstone project

光井·模宅 农村住房产业化黄石项目

光井·模宅 单体光照分析·组团院落生成 Unit Illumination Analysis · Group Courtyard Generation

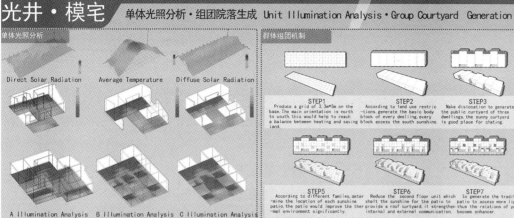

光井·模宅 主动与被动太阳能利用策略 Active and Passive Solar Energy Strategy

Heat Absorption

The south facade absorb the solar energy mainly, so most of bedrooms and living space are arranged in the south as much as possible. As the center of solar absorption, the solar patio is arraged to access the main living space as much as possible. The courtyards in different height provide the posibility of above.

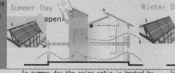

Heat Preservation

In winter the cold wind from the north would be an important factor to cause heat loss, so the stairs, bathroom and other auxiliary rooms are arranged in the north to work as buffer space. The windows to north would be narrow also. One another this strategy also suit the present rural life habit.

Through-draught

In order to improve indoor air quality, the routes of the through-draught have been rationally arranged to go through living space and bedrooms as much as possible. The entrance in the north is narrow and the exit in the south is wide, this would do good to speed up the through-draught and reduce heat loss.

solar patio_interior1
solar patio_interior2
solar patio_interior3

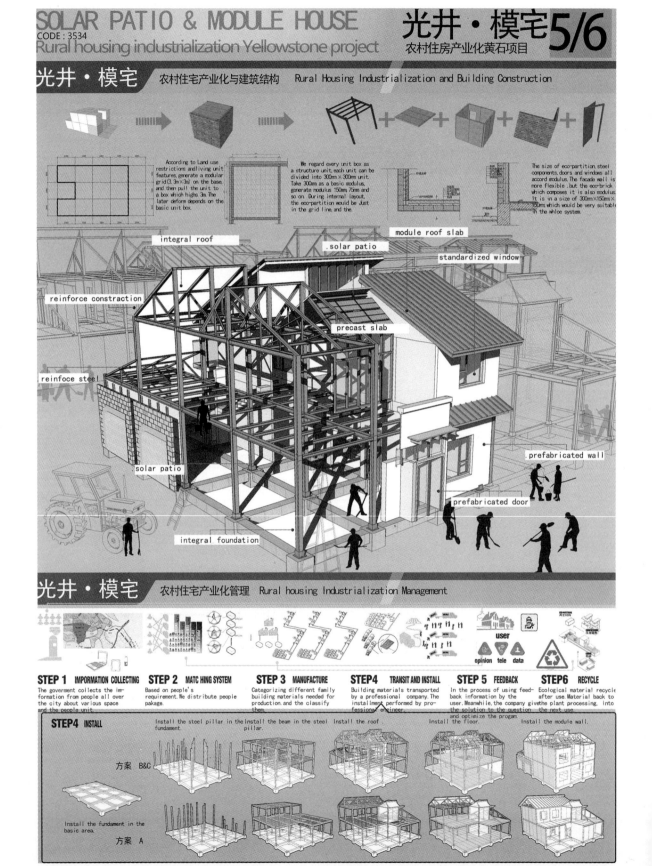

SOLAR PATIO & MODULE HOUSE
CODE : 3534
Rural housing industrialization Yellowstone project

光井·模宅 6/6
农村住房产业化黄石项目

2015 台达杯国际太阳能建筑设计竞赛获奖作品集

光井·模宅 产业化与建筑材料 Industrialization and Building Materials

光井·模宅 产业化经济分析 Industrialization of Economic Analysis

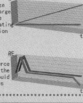

Compare with normal situation

Schedule
At the beginning of the progress, the construction of the industrialiaze residential is fater than the normal ones becase of the pre-fabricated construction methods, this would reduce the total time payed. Then, in the more meticulous progress, the speed would slow down as the normal situation.

Expenses for Energy
At the beginning of the life cycle of industrialize residential, the expenses for energy would be more than the normal ones, mostly due to the solar project and Heating project, Then, in about 6 years, the eatra expenses for extre equipments would be offseted by the energy saving on them.

Energy Consumption
In the whole life cycle of the industrialize residential, the solar project would supply a large part of energy demand in everyday life, and the mathane praject would do good to reduce the heating expense. In results, the total energy comsumption would increase slowly.

Resource Consumption
At the begining of the life cycle of industrialize residential and during use, the resource comsumption is similar as the nomal ones. But at the recovery progress, much of the main structure could be reused, especially the steel part. This proves industrialization do good to Ecology.

光井·模宅 产业化住品设计 Industrialization Building Product Design

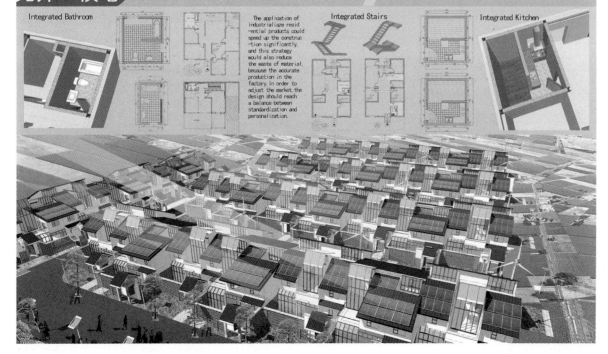

Integrated Bathroom   Integrated Stairs   Integrated Kitchen

The application of industrializae residential products could speed up the construction significantly, and this strategy would also reduce the waste of material, because the accurate production in the factory. In order to adjust the market, the design should reach a balance between standardization and personalization.

综合奖·优秀奖
General Prize Awarded·
Honorable Mention Prize

注 册 号：3631
项目名称：光临涵舍（黄石）
　　　　　Light through Country House
　　　　　(Huangshi)
作　　者：杜相、王强、胡达、
　　　　　吴昊、季京京、李欣
参赛单位：南京工业大学、合肥工业大学

光临涵舍
Light through country house　黄石住宅公园项目太阳能住宅设计　01

■Site Analysis

Situation plan

Huangshi city is located in the southeast of Hubei Province, the South Bank of the middle reaches of the Yangtze river. Huangshi City, cross 29 degrees north latitude 30 degrees 15 'East' ~30, 114 degrees ~115 degrees 30 '31', located in Chinese in southeastern Hubei Province is an important city of Wuhan Economic circle.

The competition project is located in the Huangshi residential park. It is outside the window residential industrialization of Huangshi, the overall layout for the integrated demonstration in hot summer and cold winter area of our country urban residential collection, low rise residential and rural housing to provide cultural heritage and the application of new technology.

Climate

	max	min	average
sunshine hours	2280.9	1666.4	2050.3
temperature	29.2	3.9	17
humidity	53%	46%	48%
wind speed	4.25	1.76	2.17

Current Situation

The Local Dwellings Analysis

Culture Analysis

鄂南民居分布位置

鄂南，是指我国中部内陆省份湖北省的南部地区，可以在空间上和文化上对其概念与范围作出界定。狭义地看，指现今的咸宁市及其周边范围。在鄂南地区周围还存在有多支各具地方特色的亚文化，如徽州文化、赣江文化、湖湘文化，以及峡江文化。鄂南的地域文化处于四支亚文化的边缘，其呈现出被边缘中的特点，在同一性中包含特殊性，在特殊性中体现同一性。

设计说明：
本设计从农村住宅原型出发，采用太阳能技术，达到环保节能要求，符合黄石当地气候要求。

The design prototype starting from rural housing, use solar technology to achieve green energy requirements, meeting the requirements of the local climate of HuangShi.

Dwellings Analysis 1

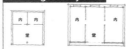

"一堂两室"示意图

在中国传统民居建筑的平面构成中，有一种基本型，称为"一堂两室"，其特征是"一明两暗"，这种简单的民居形式是其他类型平面格局的基础，可以说是其他所有平面形制的原型。

In the plane Chinese traditional residential buildings in the form, there is a basic type, called "a two", its characteristic is "a next two dark", this simple form of residential areas is the basis for other types of planar pattern, can be said to be the prototype of all the other plane shapes.

Dwellings Analysis 2

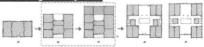

"一堂两室"向天井院演化

天井院的围合形式实际可看做是"一明两暗"式的进一步演化。它是在堂屋背后院子的后方再加上一排房屋，使原本半围合的院子成为四面围合的井院空间。形成具有鄂南地方特色的平面样式。

Courtyard enclosed form can be regarded as "a further evolution of next two dark" type.It is in the room behind the rear yard plus a row of houses, so that the original semi Wai courtyard become surrounded by Wai well courtyard space. The formation of plane style with local characteristics of South Hubei province.

Dwellings Analysis 3

鄂南典型民居天井院与马头墙

■Perspective

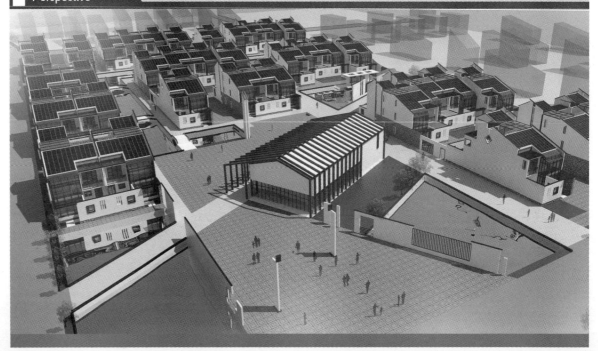

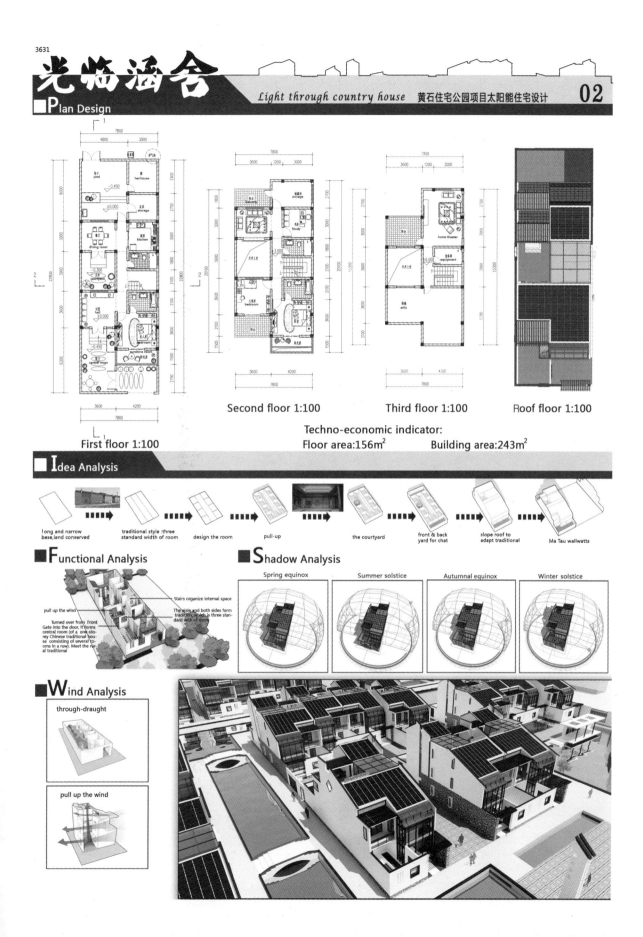

光临涵舍
Light through country house — 黄石住宅公园项目太阳能住宅设计 03

■ Section Analysizs

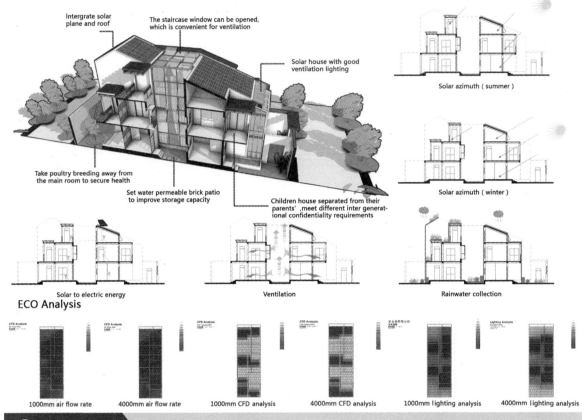

■ Section Design

1-1 Section 1:100　　2-2 Section 1:100

■ Facade Design

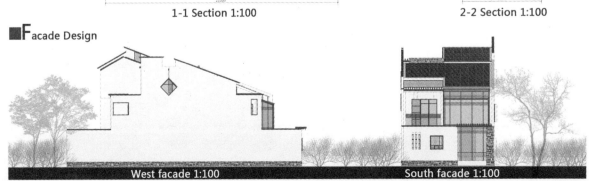

West facade 1:100　　South facade 1:100

光临涵舍
Light through country house
黄石住宅公园项目太阳能住宅设计
04

■ Technique Analysis

Solar panel
split
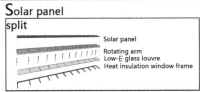

structure
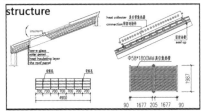

photovoltaic system
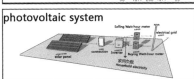

option
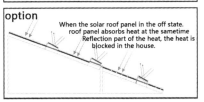
When the solar roof panel in the off state. roof panel absorbs heat at the sametime Reflection part of the heat, the heat is blocked in the house.

Total analysis
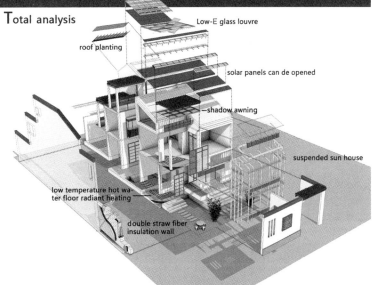
- Low-E glass louvre
- roof planting
- solar panels can de opened
- shadow awning
- suspended sun house
- low temperature hot water floor radiant heating
- double straw fiber insulation wall

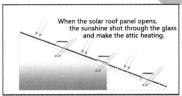
When the solar roof panel opens, the sunshine shot through the glass and make the attic heating.

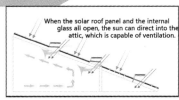
When the solar roof panel and the internal glass all open, the sun can direct into the attic, which is capable of ventilation.

Solar house
principle(summer)
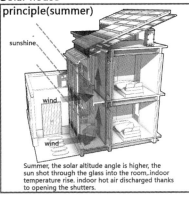
Summer, the solar altitude angle is higher, the sun shot through the glass into the room, indoor temperature rise. indoor hot air discharged thanks to opening the shutters.

principle(winter)
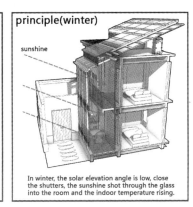
In winter, the solar elevation angle is low, close the shutters, the sunshine shot through the glass into the room and the indoor temperature rising.

structure
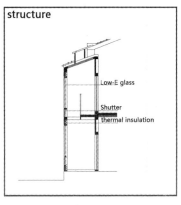
- Low-E glass
- Shutter
- thermal insulation

Trombe wall
air-out
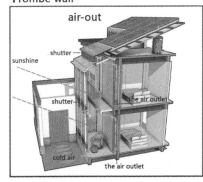

air-in
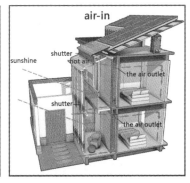

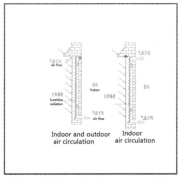
Indoor and outdoor air circulation | Indoor air circulation

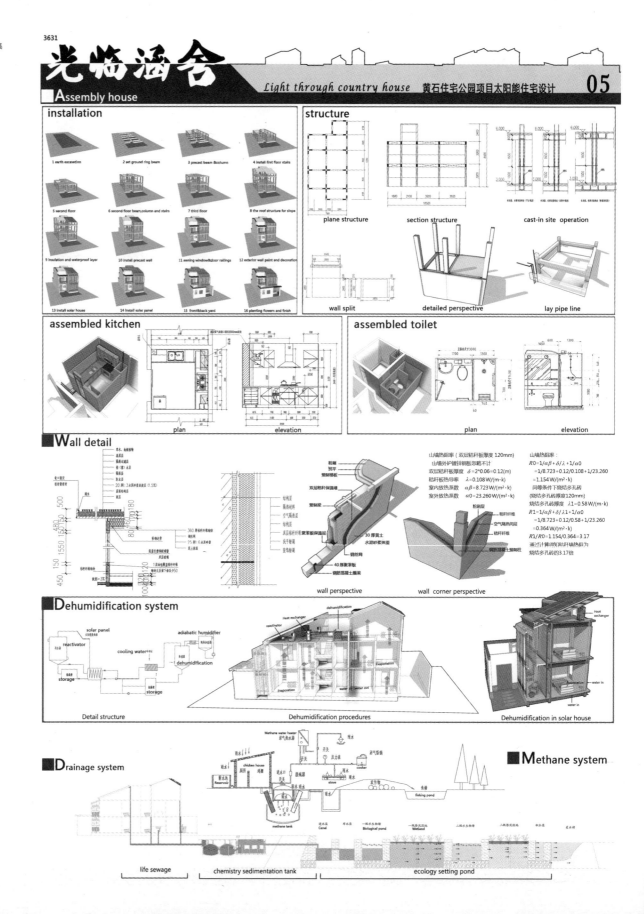

光临涵舍

Light through country house — 黄石住宅公园项目太阳能住宅设计

06

■ **L**ocal row house　　■ **L**ocal alley

link mode
A enclose the yard
B parallel and cross
C joint

Courtyard South Hubei traditional houses is constitute a considerable momentum with the scale of the "big house" There are three to five or more; tianjingyuan transverse linkcalled "joint".

Beginning and end of street intersection is a street, and intermediate nodes to another street lane. Most of these intersections are the "T" - shaped

this plan design cantain many "T" -shaped alley which remind people of old village.

■ **S**ite Plan　　■ **A**daptability

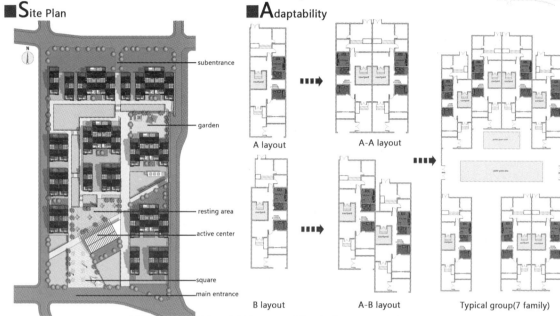

■ **F**acade

plan facade 1:250

综合奖·优秀奖
General Prize Awarded · Honorable Mention Prize

注 册 号：3723
项目名称：揽境（黄石）
Embrace the Environment (Huangshi)
作 者：尹 欣、加晶晶、仇朝兵、
陆文蕙、方东亚、侯丹蕾
参赛单位：石家庄铁道大学、河北建筑工程学院

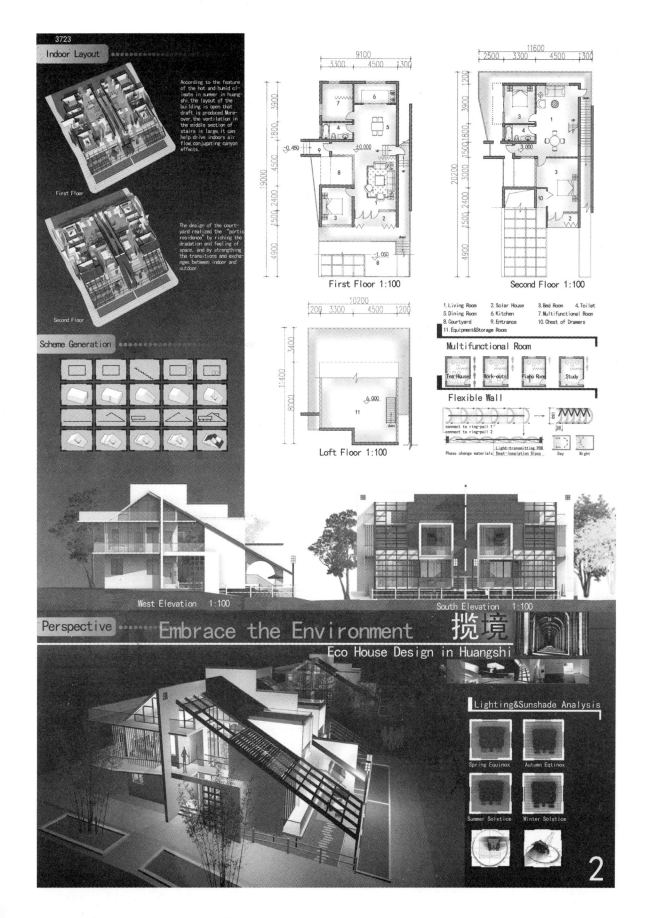

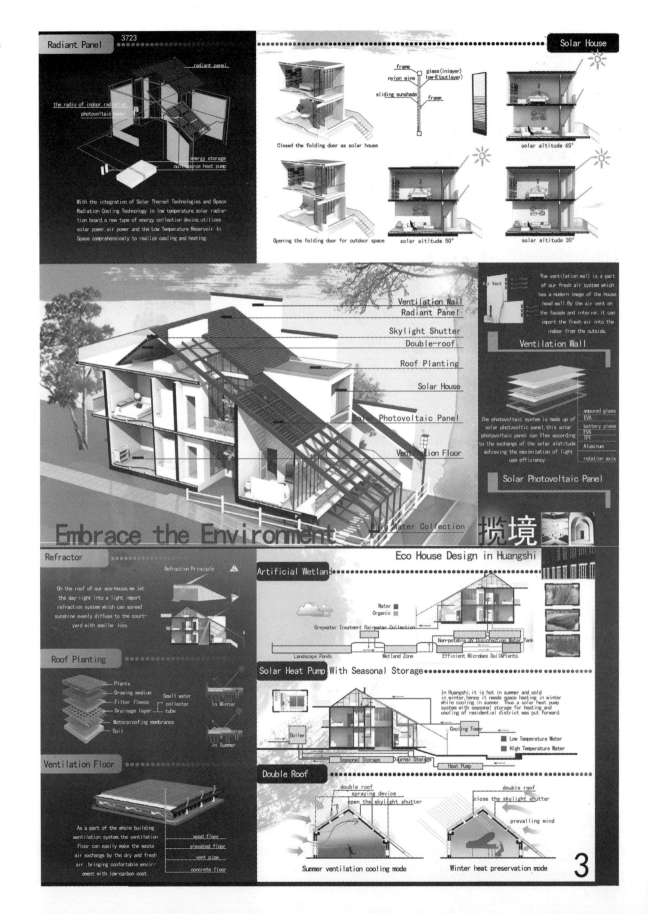

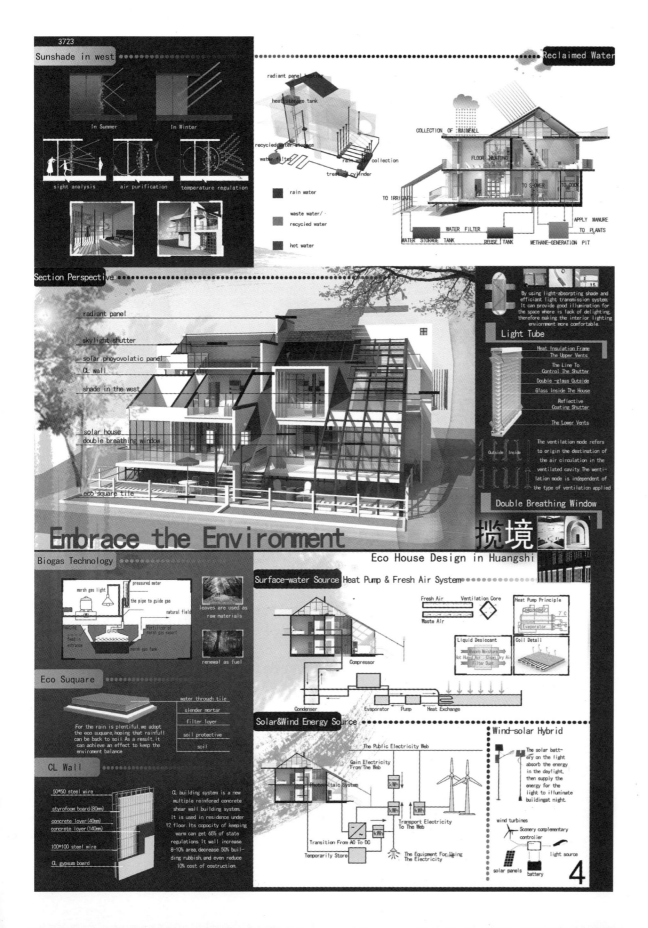

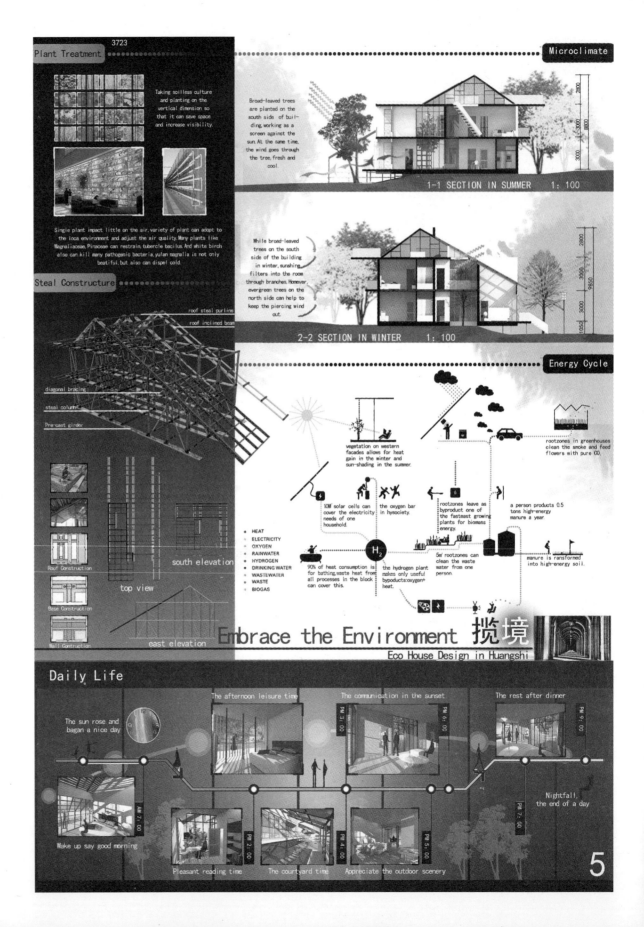

综合奖·优秀奖
General Prize Awarded·
Honorable Mention Prize

注 册 号：3826
项目名称：乡人居所（黄石）
　　　　　Villager Residence (Huangshi)
作　 者：石 成、丁 和
参赛单位：暨南大学、山东建筑大学

乡人居所 Villager residence

Econo-technical indicators

Building A

building area	206.94m²
floor area ratio	1.38
building density	84.77%
usable area	154.78m²
usable rate	74.79%
greening area	52.69m²
greening rate	25.46%

Building B

building area	205.74m²
floor area ratio	1.37
building density	76.49%
usable area	170.08m²
usable rate	82.67%
greening area	37.52m²
greening rate	18.24%

Active solar design

gary space

natural ventilation

garden lighting

passive solar sunroom

A Plan

1F Plan 1:100   2F Plan 1:100

B Plan

1F Plan 1:100   2F Plan 1:100

Block analysis

A: traditional block | malocclusion | occlusion ; sink | virtual space | model

B: traditional block | sink ; forming a courtyard | upraise | virtual space | model

west elevation 1:100   east elevation 1:100   south elevation 1:100   north elevation 1:100

乡人居所 Villager residence

Model analysis

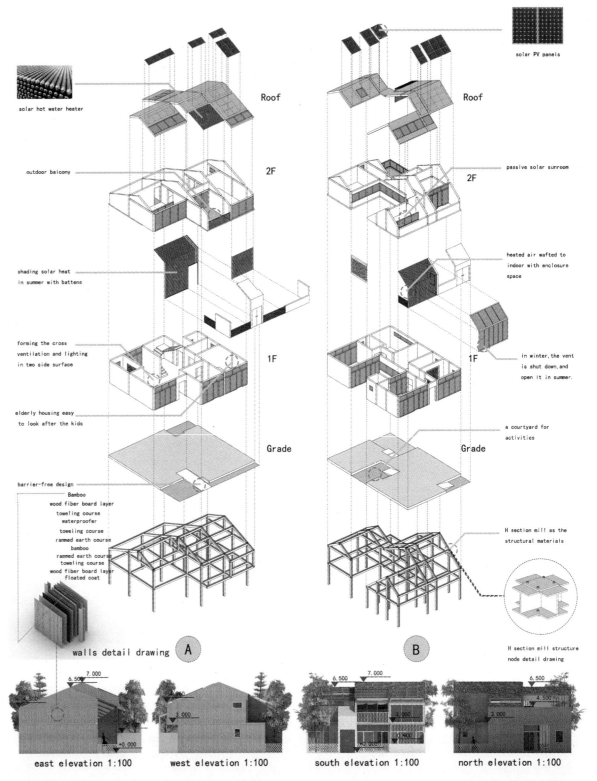

乡人居所 Villager residence

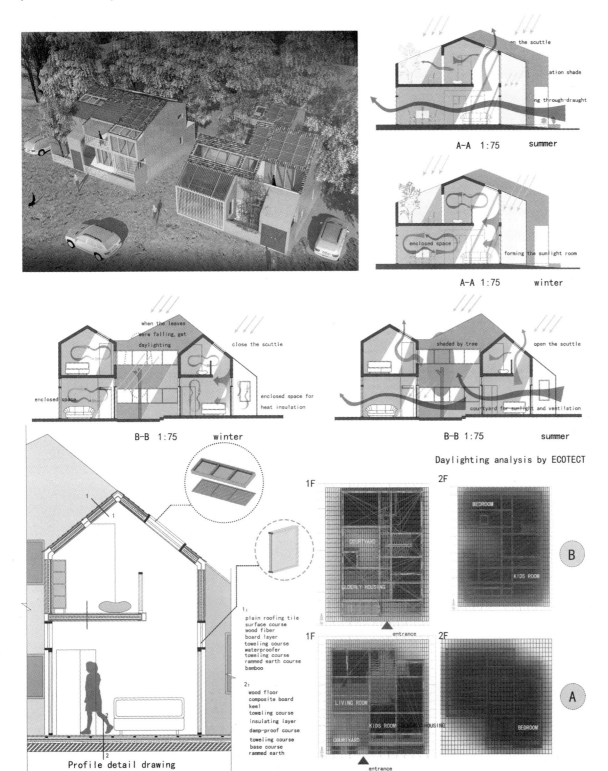

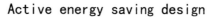

乡人居所 Villager residence

Active energy saving design
Heat pump heat supply

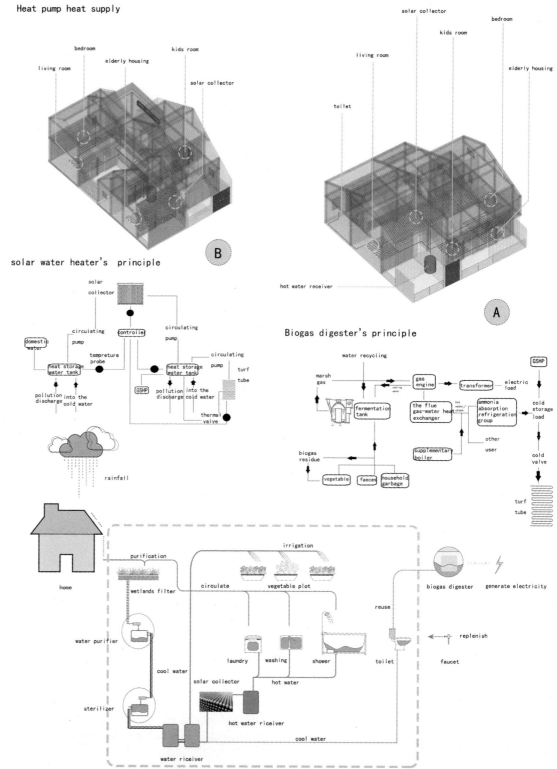

乡人居所 Villager residence

Housing cluster analysis

Site-plan analysis

public building and plantation | elderly housing | multi-family housing | afforest | plan result

The community running mode analysis

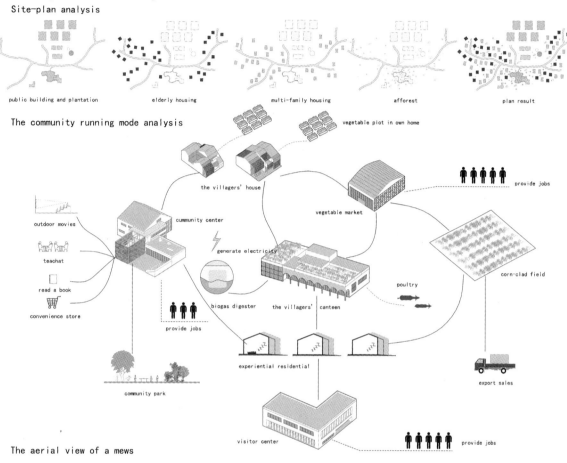

The aerial view of a mews

综合奖·优秀奖
General Prize Awarded·
Honorable Mention Prize

注 册 号：3972
项目名称：竹·光·园·间（黄石）
　　　　　Bamboo·Light·Courtyard·
　　　　　Space (Huangshi)
作　　者：高深、张振、贾峰、
　　　　　张信雅、陈晓东、程子瑜、
　　　　　杨镇、高晶、张建凤
参赛单位：吉林建筑大学、东北石油大学

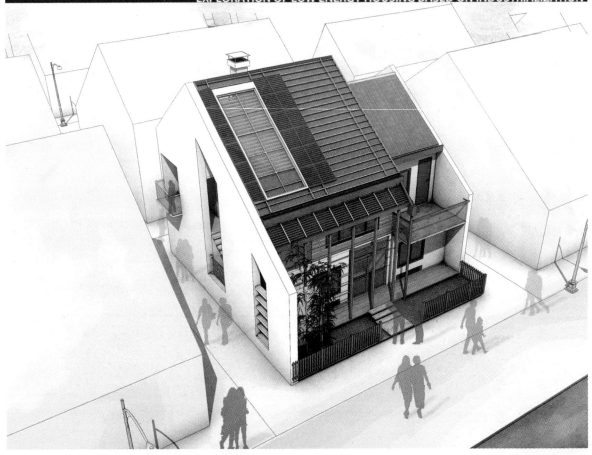

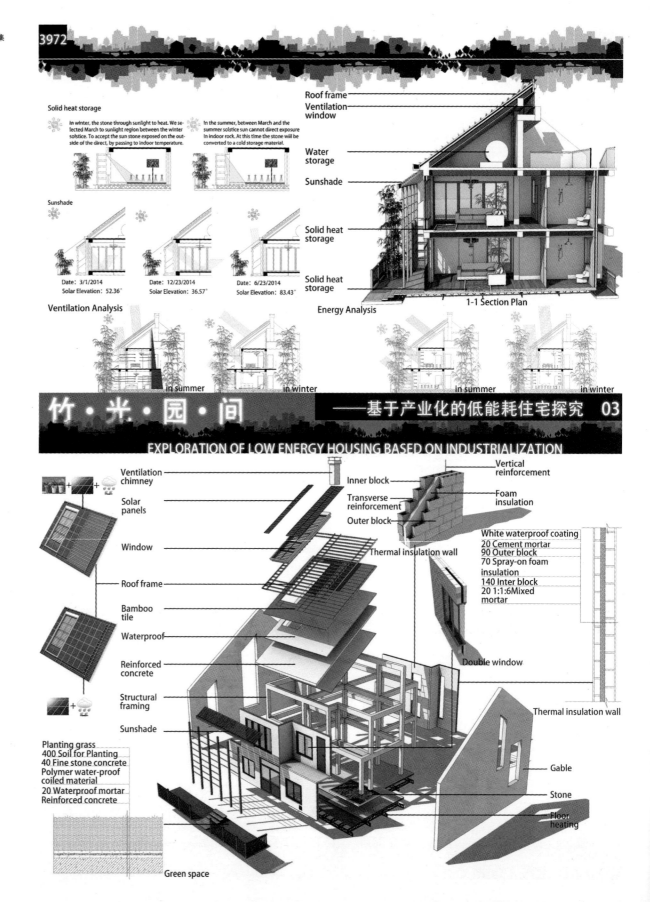

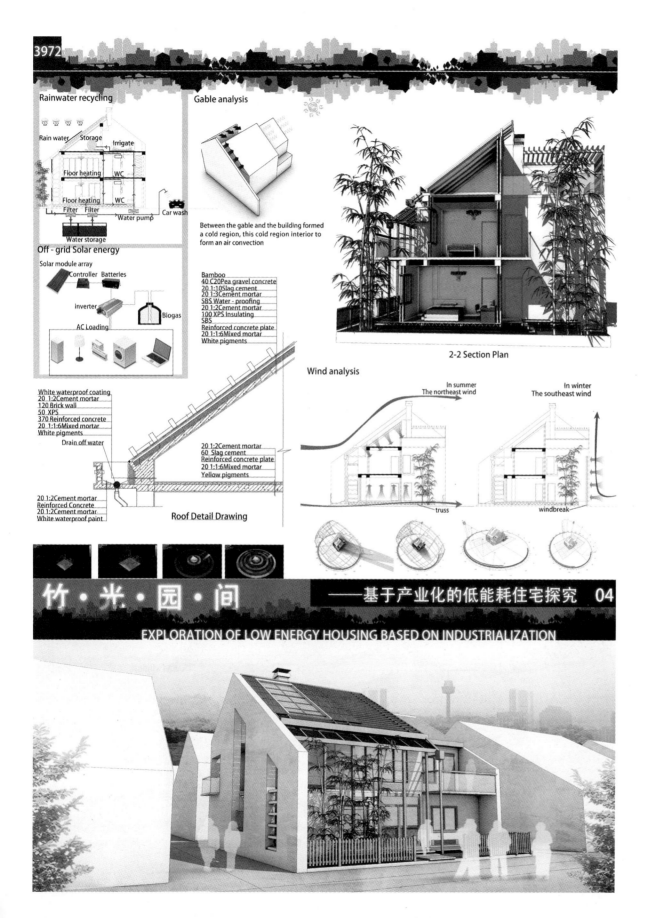

综合奖·优秀奖
General Prize Awarded · Honorable Mention Prize

注 册 号：3985
项目名称：家×宅（黄石）
　　　　　Home×House (Huangshi)
作　　者：庄梓涛、肖泽恒、余光鑫、
　　　　　张贵彬、王良亮、李张君、
　　　　　刘智伟、李玥
参赛单位：华南理工大学建筑学院青年设
　　　　　计工作室

家×宅 HOME X HOUSE

壹 | CONCEPT

CONCEPT ANALYSIS

? UNCERTAINTY OF HOME—TRADITIONAL RURAL CONDITION ANALYSIS
Left behind children, elderly person of no family, labour migration and rural construction disorderly are the main contradictory issues in Hubei village.

DEVELOPMENT OF HOUSE—FRUITION OF MODERN HOUSING TECHNOLOGY
Green technology can use limited resource to provide high living quality. Industrialization and prefabricate assembly technology can cut down construction cost, improve building quality and control disorder construction.

HOME X HOUSE COMBINATION—RURAL IMAGE COMBINE WITH MODERN TECHNOLOGY
Traditional patio integrates with modern sun room. Courtyard infiltrates indoor space. Traditional double roof combines with modern structure.

SITE ANALYSIS

Huanshi, Hubei
TRANSPORTATION: Transportation is convenient for local economic development.
RESOURCE: Resource is rich in eletricity, steel, cement, plaster, sand and stone.
GEOGRAPHY: Terrain is mainly plain.
CLIMATE: Hot-summer and cold-winter; wet and humid

设计说明：

家，是人记忆的载体，象征农村从古生长传承的文化；宅，指人居住的场所，代表城市化进程中住宅产业化技术发展的成果。现在农村正处于社会转型期，属于农民的"家"因为文化与技术发展的不同步导致断层的出现，并且演变出很多的社会问题。而此时我们希望可以通过立足农民的生活，以保留农村本身的记忆为前提，结合产业化技术和绿色技术的成果，通过合理的建筑设计，构建得以在当代继续传承农村文脉的场所。

HOME is where memories take place, the carrier of rural culture. HOUSE is the place where people live, it stands for the fruition of city and modern technology progress. Chinese countryside is at the point of social changing. HOME is faced with the conflict of tradition culture and technology development, which result in many social problems. We wish to base our design on the life of peasant, to preserve memories in village and meanwhile combine it with modern industrial and green technology. By architectural design, we

DESIGN CLUE:

贰 SPACE DESIGN — Combine the traditional space mode with modern functions to create a comfortable living condition

叁 MENU DESIGN — The menu allows residents to choose their house accordingly, meeting the requirement of sustainable development

肆 PASSIVE DESIGN — Use the high efficiency passive technology to create good indoor environment and reduce the consumption

伍 INDUSTRIALIZATION DESIGN — Building industrialization is good for controlling construction quality, reducing construction consumption and promting local employment

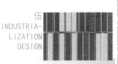
陆 SYSTEM DESIGN — Good system design helps to save the energy and room, which is easier to maintenance

家×宅 HOME X HOUSE   贰 | SPACE DESIGN

DESIGN BACKGROUND

1. Pineline in rural houses are not compact enough and often cause leakage.
2. Patio is traditional space in Hubei, where many activities carried out around it.
3. Courtyard is important in rural life and can be use flexibly.
4. Public space is important is traditional villages which is now dying.

DESIGN STRATEGY

1. Space with some functions are arranged closely for compacting pineline and structure modulization.
2. A patio is set between the main entrance and indoor space as a transition, which is inherited from traditional space order. Back entrance and side entrance is convenient for daily use.
3. Different ways of houses joining creates diversity in village space, which also satisfying the need of public and neigborhood, renewing cordial feeling in rural public lives.

ARCHITECTURE VOLUME GENERATION | FUNCTION MODULE ANALYSIS | TRAVEL PATH ANALYSIS

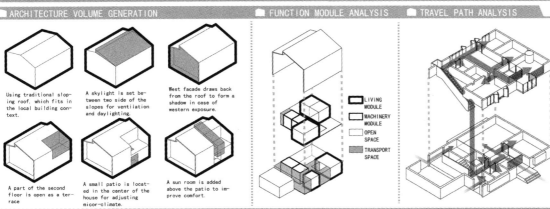

- Using traditional sloping roof, which fits in the local building context.
- A skylight is set between two side of the slopes for ventilation and daylighting.
- West facade draws back from the roof to form a shadow in case of western exposure.
- A part of the second floor is open as a terrace.
- A small patio is located in the center of the house for adjusting micor-climate.
- A sun room is added above the patio to improve comfort.

LIVING MODULE
MACHINERY MODULE
OPEN SPACE
TRANSPORT SPACE

FIRST FLOOR PLAN 1:100

SECOND FLOOR PLAN 1:100

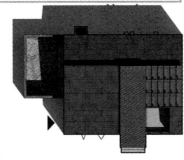
ROOF PLAN 1:100

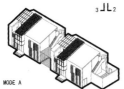
MODE A
Houses with the direction can be joined without disturbing one another. A small transition space is formed between every two houses.

MODE B
House can also be joined symmetrically.

MODE C
The side with the courtyard can be joined so that a bigger publick space is created between houses for communican and daily use of the neigborhood.

家x宅 HOME X HOUSE   叁|MENU DESIGN

3985

MENU DESIGN ANALYSIS

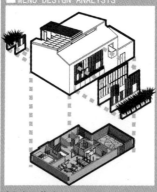

DESIGN BACKGROUND

1. Concrete and brick structures are widely used in rural area, which lacks of flexibility.

2. Some villages are developing tourism, where intermixing of commerce and residence appears.

DESIGN STRATEGY

1. Providing MENU for different needs of residents.
2. The building consists of different components that can be adjusted according to the MENU.
3. Different components bring different openness to the first floor, which suits the needs of various life styles.

Living Mode
Living mode is perfect for a normal family. Windows facing the streets on the first floor are raised above sight level to ensure privacy. There's a room for the elderly on the first floor, which is convenient for the seniority, and also relatively divides the lives of the old and the young to avoid generation gap. Meanwhile they can enjoy kinship in the dining room and living room.

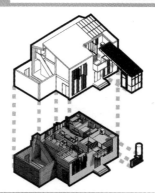 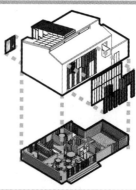 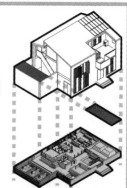

Rental Estate
The first floor can be transformed into a rental or youth hotel by adding partitions in living room and elderly room of the first floor. For the semi-open stair on the west can completely divide the first and second floor, the host can run a rental estate as well as maintain his own private space on the second floor.

Sun room and collector tubes
Users can decide whether to add sun room and collector tubes on the house. Sun room can improve the micro-climate within the house by absorbing radiations in winter to gain heat and fully opened in summer to improve ventilation. Collector tubes are integrated with handrail as part of the house.

Business mode
Using full height glazing, the rooms can transformed to public spaces for business. The first floor living room can be changed into a restaurant while the elderly room can be changed into a store for their directly facing the street. All these can be selected from the building menu.

Production and living
Adding cover and partitions for the backyard will allow the host to breed fowl and live-stock. The elderly room, which is directly below the terrace, can also be transformed into a barn for storing food and fodder. Cereal can be dry in the sun on second floor and directly move to the barn below if we add a transit path.

SECTION 1-1 1:100   SECTION 2-2 1:100   SECTION 3-3 1:100

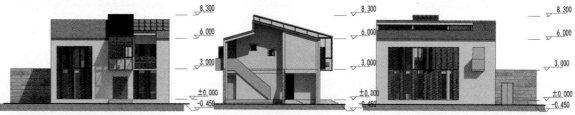

SOUTH ELEVATION 1:100   WEST ELEVATION 1:100   NOUTH ELEVATION 1:100

家x宅 HOME X HOUSE 肆 | PASSIVE DESIGN

DESIGN BACKGROUND

1. Traditional folk houses use many passive design to create comfortable living enviroment.

2. Hubei is in hot-summer and cold-winter zone. Cooling and heating both need to be considered in design.

3. Tanglong gate can improve ventilation and ensure security at the same tiem, can be open at summer night for cooling

4. Sun room benefits the house in winter while gain much heat in summer.

DESIGN STRATEGY

1. Learn passive design from folk house such as patio.

2. Using louver and openable glazing combined with sun room to improve its condition in summer.

3. Combining Tanglong gate with wooden allow user to control ventilation for different cases.

4. Using adjustable louver to deal with different weather condition, giving the control right to the user.

5. Using sitmulation software to situmulate passive design and adjust the relatively best solution for the house.

PASSIVE DESIGN ANALYSIS

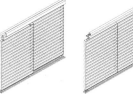

SLIDING LOUVERS (CLOSED)   FOLDING LOUVERS (CLOSED)

SLIDING LOUVERS (OPENED)   FOLDING LOUVERS (OPENED)

SLIDING LOUVERS DETAIL (1F)   FOLDING LOUVERS DETAIL (2F)

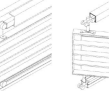

Sunlight fully penetrates the louvers in winter to heat the indoors.

Sunlight are blocked in summer to avoid heating the sun room. Opening windows on the sun room can improve ventilation.

Horizontal louvers on the south can block sunlight with 60° solar altitude (From Apr. to Oct.) and penetrate sunlight from Nov. to Mar..

Folding louvers on the north can be use as vertical shading, blocking sunlight before 10:00 am. and after 4:00 pm. in summer.

The gate consists of two parts: the wooden solid door and the sliding fence. Usually both the door and fence are opened for daily use.

The wooden solid door can be closed to block cold air in winter.

In hot weather, the solid door can be opened while the sliding remain closed, which allow air flows while ensuring security.

SIMULATION

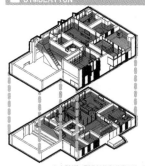

LIGHT ENVIRONMENT SIMULATION

Ventilation is important for improving comforts in Huangshi, Hubei because of humidity. Prevailing wind in Huangshi comes from southeast with an average speed of 4m/s. The house's plan and sections are adjusted for satisfying ventilation.

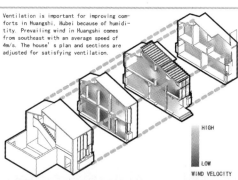

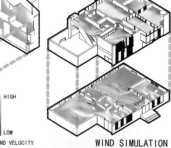

WIND SIMULATION

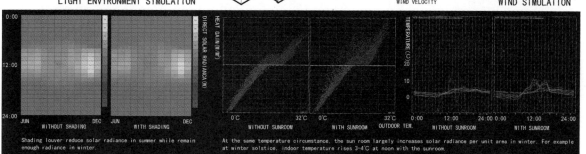

Shading louver reduce solar radiance in summer while remain enough radiance in winter.

At the same temperature circumstance, the sun room largely increases solar radiance per unit area in winter. For example at winter solstice, indoor temperature rises 3-4°C at noon with the sunroom.

家×宅 HOME X HOUSE | 伍 INDUSTRIALIZATION DESIGN

DESIGN STRATEGY

1. Searching for a construction way that can fit in both large quantity and good quality

2. Using economical structure to guarantee structural strength and insulation while being affordable for peasants

3. Using local materials to activate the local industry and economy.

DESIGN BACKGROUND

1. Most rural residences in Hubei are built by peasants themselves, which lack of earthquake fortification and qualified building materials.

2. Brick-concrete structure is mainly used in rural houses, which has low construction efficiency and serious environmental pollution.

3. Huangshi has rich resources such as steel, cement, sand and stone, making it possible to build steel, reinforced concrete and masonry structure.

4. To cut down costs, peasant only choose cheap material and low-tech measure in constructions.

STRUCTURE FORM SELECTION ANALYSIS

Based on local residence conditions in Huangshi, we use PRE-CAST PRESTRESSED ASSEMBLED MONOLITHIC CONCRETE STRUCTURE. Structural element are prefabricated in factories and shipped to construction site, with partial cast-in-place concrete to strengthen over stiffness. Precast concrete can reduce construction waste, shorten construction period, and promote the development of housing industrialization.

PRODUCTION COST (YUAN/m²)	ANNUAL USAGE COST (YUAN/m²)	ANNUAL MAINTAINING COST (YUAN/m²)	WHOLE LIFECYCLE COST (YUAN/m²)
2000 / 1400 / 1000	52 / 60 / 48	25 / 18 / 10	5250 / 5200 / 4500

Though compared to brick-concrete structure, PC structure has higher initial construction cost, it has lower usage and maintenance cost during the future decades of living. Compared to steel structure, PC has overall lower cost and comparable quality. PC is the best choice in Hubei where cement and sand resources are rich.

PC WALL   LIGHTWEIGHT WALL   BRICK WALL

Lightweight wall has overall good performances but higher price which makes it low cost performance in rural are. PC wall is cheaper while having evenly good performances.

T - THERMAL INSULATION
W - WALL THICKNESS
I - INTENSITY
F - FIRE-RESISTANCE
S - SOUND INSULATION

INDUSTRIALIZATION AND INTEGRATION

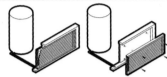

Beams, columns and floors are prefabricated in factories. PRECAST CONCRETE SANDWICH PANELS are used as envelope. The module of the house is 1200mm according to the panels. Doors and windows are produced along with the walls in the factory for rapid assembly.

Solar heat-collecting tubes are integrated with handrail on the terrace as part of the building components without affecting the integral facade.

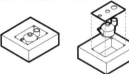
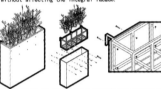

Louvers on the windows provide shade and animate elevations. As industrial components, they can be chosen and purchased according to the needs of the owner.

The flower-stands are produced in module, whose size are decided according to the window area for easy installation and removal.

Biogas digester integration can improve gas tightness and increase construction efficiency. It can be shared by the neighborhood.

PV panels are prefabricated in factories. The panel's joists are connected to concrete roof by bolting to the pre-buried steel angle.

STRUCTURE AND CONSTRUCTION ANALYSIS   JOINTS DETAILS

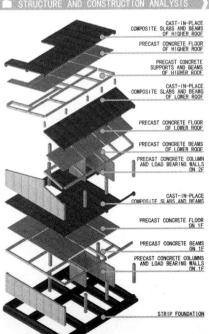

- CAST-IN-PLACE COMPOSITE SLABS AND BEAMS OF HIGHER ROOF
- PRECAST CONCRETE FLOOR OF HIGHER ROOF
- PRECAST CONCRETE SUPPORTS AND BEAMS OF HIGHER ROOF
- CAST-IN-PLACE COMPOSITE SLABS AND BEAMS OF LOWER ROOF
- PRECAST CONCRETE FLOOR OF LOWER ROOF
- PRECAST CONCRETE BEAMS OF LOWER ROOF
- PRECAST CONCRETE COLUMN AND LOAD BEARING WALLS ON 2F
- CAST-IN-PLACE COMPOSITE SLABS AND BEAMS
- PRECAST CONCRETE FLOOR ON 1F
- PRECAST CONCRETE BEAMS ON 1F
- PRECAST CONCRETE COLUMNS AND LOAD BEARING WALLS ON 1F
- STRIP FOUNDATION

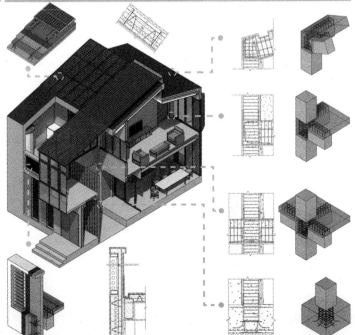

家x宅 HOME X HOUSE — 陆 | SYSTEM DESIGN

DESIGN BACKGROUND

1. In 2014, Huangshi, Hubei has put forward solar energy projects with installed capacity of 300 MW to uphold solar power generation industry.

2. Hubei is called The Province of Thousands of Lakes. Its underground water sources and ground regenerators are steadily within 15~20°C.

In 2009, Hubei passed Hubei Building Energy Efficiency Regulations, encouraging building projects to use Ground Source Heat Pump.

In 2013, National Energy Administration requires Hubei to constitute biomass renewable energy system, where the rural areas have rich biomass energy resources.

DESIGN STRAGETY

Integrating solar power with biomass energy.

Using political-encouraged techniques.

1+1>2
Integrating technologies at system level.

Using high cost efficient techonology.

SYSTEM DESIGN

SOLAR ENERGY GENERATION SYSTEM
Technological Innovation:
1. Building Integrated Photovoltaics. Pre-tinstall PV components when prefabricating the roof.
2. The house's PV system is connected to civil power grid, which can sell extra energy to the civil grid. SOLAR ENERGY GENERATION SYSTEM

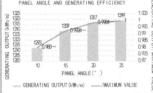

PANEL ANGLE AND GENERATING EFFICIENCY

Adjust the panel's angle according the generating efficiency and aesthetics views

WATER HEATING SYSTEM
Technological Innovation:
1. Hot water system connected with biogas digester to keep steady temperature.
2. The radiant bed receive heat resource from heating and cooling system to radiate heat and cooling capacity to improve indoor comfort.
3. Using GSHP (Ground Source Heat Pump) is supported by the local policy.
4. Using dehumidifier helps to cope with the humidity in Hubei.

H-HEAT PRESERVATION
F-FROST RESISTING
P-PRESSURE-BEARING
R-RAY TRACING
C-COST S-SECURITY

FLAT PLATE COLLECTOR / U-BENT ECAVUATED COLLECTOR TUBE / GLASS-METAL COLLECTOR TUBE / STRAIGHT-FLOW EVACUATED COLLECTOR TUBE

P-POLICIAL UHOLD
F-Fee
F-FUNCTION
I-INITIAL COST
E-ENERGY CONSUMPTION

ELECTRIC ENERGY / VARIBALE WATER TEMPERATURE / WATER SOURCE HEAT PUMP / GROUND SOURCE HEAT PUMP

BIOGAS DIGESTER
Technological Innovation:
1. Hot water system connected with biogas digester to keep steady temperature.
2. Industrial production of biogas digester to improve air tightness.

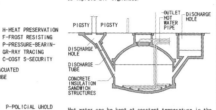

Hot water can be kept at constant temperature in the digester. Polyurethane sandwich in concrete component help to improve insulation and improve digester's efficiency.

SYSTEM DIAGRAM

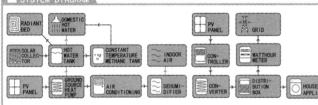

DIAGRAM OF WATER HEATING SYSTEM

In summer, GSHP is used for cooling and auxiliary heating hot water provided by solar thermal collector. The hot water can be kept at constant temperature in the biogas digester and delivered to radiant bed for heating the house. Domestc hot water is provided at this circulation.
To cope with humidity in Huangshi, dehumidifier is integrated the HVAC system

DIAGRAM OF GRID-CONNECTED PV SYSTEM

PV system produces energy to satisfy the house and sells extra electricity to the local power grid. It saves the cost of purchasing accumulator. Meawhile the government will provide subsidies for generating and selling extra electricity. (It takes 7 year to reclaim system cost)

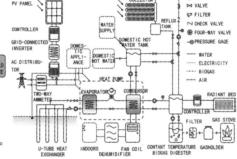

SYSTEM DETAILED DIAGRAM

SYSTEM COST

Grid-connected solar system can save the cost of storage battery. By government subsidy, energy saving and selling extra electricity, the cost can be reclaim in 7 years

Water-resource heat pump can save 1000¥ per year compared to normal AC system

LIVING SENCE

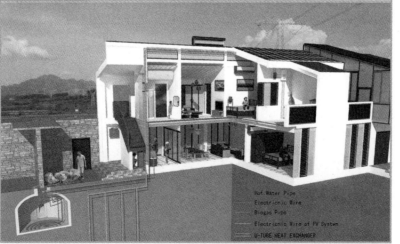

Hot Water Pipe
Electricnic Wire
Biogas Pipe
Electricnic Wire of PV System
U-TUBE HEAT EXCHANGER

综合奖・优秀奖
General Prize Awarded · Honorable Mention Prize

注 册 号：4033
项目名称：沐浴阳光（黄石）
　　　　　Green & Sunshine (Huangshi)
作　　者：沈梦乔、李佳威、李硕
参赛单位：南京工业大学、The University of Texas at Austin

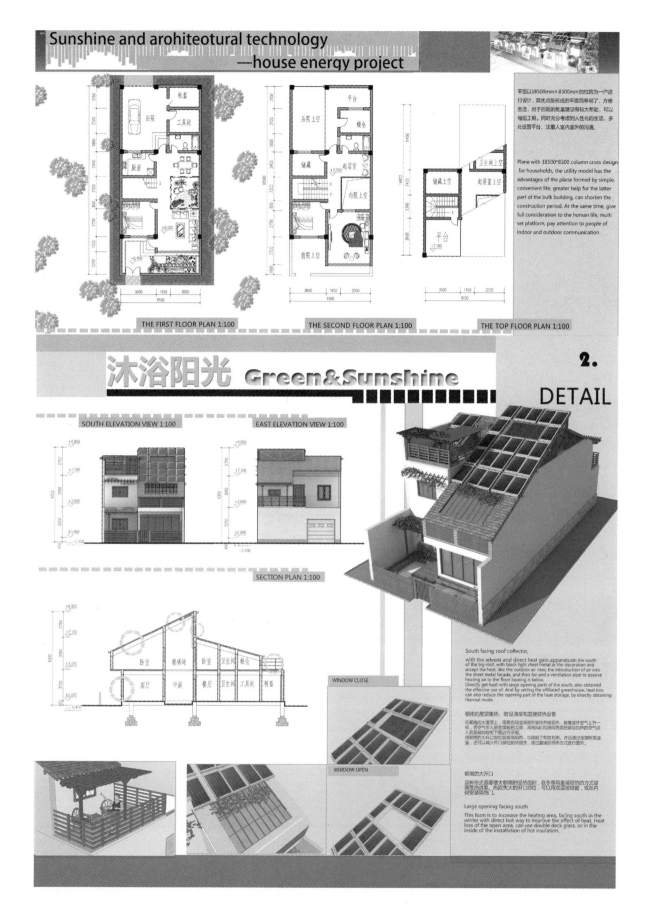

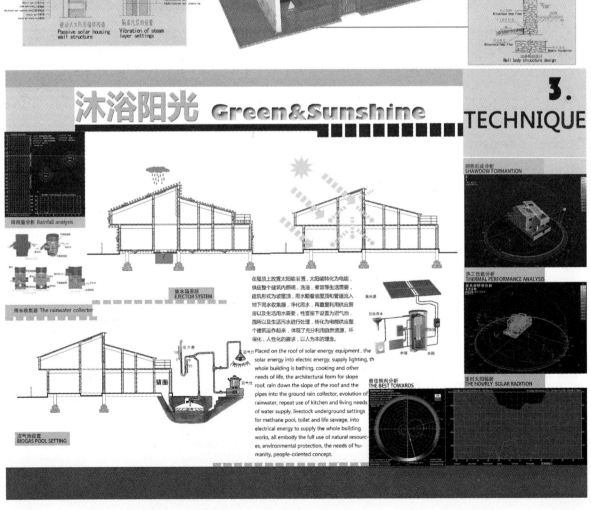

Sunshine and architectural technology
—house energy project

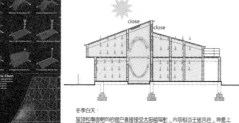

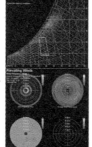

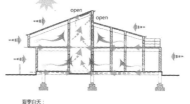

冬季白天：
屋顶和南面朝向的窗户直接接受太阳能辐射，内院相当于被风井，将最上面的窗户关闭，中间空气温度不断升高，形成一个空气循环的温室。隔热材料控制空气流进流出，白天太阳能够很好地保温。

winter day:
The roof and south towards the windows directly accept the solar radiation, the inner court is equivalent to pull a air shaft, the top window is closed, the middle air temperature rising, forming an air cycle greenhouse. Insulation materials of air flowing in and out of control. Can be very good in the daytime heat preservation.

夏季白天：
地板和屋顶都设有流通空气层，内庭顶上的窗户开启从而增加空气流通，通过屋顶单层的通风换气进行排热和防止结露，利用防潮片材防止室内湿气进入屋顶以及防止室内向外渗出。

summer day:
Floor and roof is provided with a flow of air layer on top of the inner court, the window is opened to increase air circulation, to exhaust heat and prevent condensation through ventilation roof layer, prevent moisture from entering the roof and prevent indoor leaking out using moistureproof sheet.

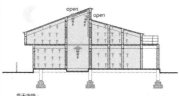

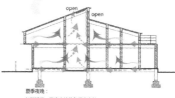

在夏季，屋顶部的气温又容易高达60~70℃。此时即使有顶部隔热，也会有大量的热气进入室内，便上层的卧室盛热难眠。为防止这种情况的出现，就要把屋顶加热的高温空气排出屋外，即采用屋顶层层顶通风会很有效果。在冬季这却可以排出湿气，起到防止结露的作用。

In the summer, the roof layer temperature was as high as 60~70 degrees Celsius. In this case, even with the roof insulation, there will be a lot of hot air into the room, the upper layer heat difficult sleep bedroom. In order to prevent the occurrence of such a situation, we should put the discharged high-temperature air roof layer outside, namely uses the roof layer ventilation will have the effect very much. In winter you can exhaust the moisture, to prevent condensation effect.

冬天夜晚：
打开由白天太阳能集热的地暖，打开内庭上方的窗户，将室内地暖的余热排出窗外，防止形成结露。

winter night:
Open the hot to warm by day solar, open the windows above the inner court, indoor warm by waste heat discharged out of the window, to prevent the formation of condensation.

夏季夜晚：
加强通风，将室内的热气带出室外。

summer night:
Strengthen the ventilation, indoor heat out of the room.

沐浴阳光 Green&Sunshine ASSIGEMENT 4.

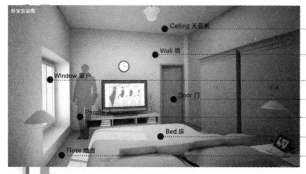

Ceiling 天花板
Wall 墙
Window 窗户
Door 门
People
Bed 床
Floor 地面

- Gypsum board 石膏板
- Concrete Block (coarse) 混凝土（粗糙）
- Glass/heavy plate 玻璃/厚
- Oak 橡木
- Adult (Per person) 成年人（每人）
- Fabric 纤维织品
- Carpet (1/8") 地毯 (3mm)

对卧室材料的选取，严重影响着卧室声环境，不同频率混响时间应在如下范围内最为适宜：
125Hz: Tr(min)=0.42s <Tr<Tr(max)=0.70s
500Hz: Tr(min)=0.41s <Tr<Tr(max)=0.53s
4000Hz: Tr(min)=0.36s <Tr<Tr(max)=0.44s

Volume 体积 (m³)	100	125	160	200	250	315	400	500	630	800	1000	1250 1600	2000 2500	3150 4000	5000
40~50	Media	0.75	0.56	0.53	0.56	0.55	0.51	0.47	0.42	0.46	0.43	0.44	0.43 0.43	0.41 0.40	0.40 0.39	
	SD	0.30	0.14	0.08	0.11	0.12	0.11	0.10	0.10	0.09	0.07	0.08	0.08 0.07	0.07 0.05	0.05 0.04	

Average Reverberation Time in bedrooms according to volume (40~50m³)
40~50m³卧室的平均混响时间
SD: Standard Deviation 标准差
Media 中间值
Source: Meta, Leonardo, and Manuel Recueri. "Simplified Method to Estimate Reverberation Time in Dwellings."

STEP2: CALCULATE THE TOTAL (ACOUSTIC) ABSORPTIVITY OF BEDROOM
步骤2: 计算卧室的总吸声系数
$A = S_1a_1 + S_2a_2 + S_3a_3 + ...$

A(125Hz)
$= \sum S(Gyp) \times a(Gyp) + \sum S(Gl) \times a(Gl) + \sum S(Co) \times a(Co) + \sum S(Ca) \times a(Ca) + \sum S(Oak) \times a(Oak) + \sum S(F) \times a(Fa) + N(P) \times a(P)$
$= 147.25 \times 0.1 + 14.64 \times 0.06 + 130.53 \times 0.36 + 79.82 \times 0.15 + 117.33 \times 0.58 + 95.8 \times 0.07 + 1 \times 2.5$
≈ 151.82 sabin

A(500Hz)
$= \sum S(Gyp) \times a(Gyp) + \sum S(Gl) \times a(Gl) + \sum S(Co) \times a(Co) + \sum S(Ca) \times a(Ca) + \sum S(Oak) \times a(Oak) + \sum S(F) \times a(Fa) + N(P) \times a(P)$
$= 147.25 \times 0.05 + 14.64 \times 0.03 + 130.53 \times 0.31 + 79.82 \times 0.40 + 117.33 \times 0.07 + 95.8 \times 0.49 + 1 \times 4.2$
≈ 139.55 sabin

A(4000Hz)
$= \sum S(Gyp) \times a(Gyp) + \sum S(Gl) \times a(Gl) + \sum S(Co) \times a(Co) + \sum S(Ca) \times a(Ca) + \sum S(Oak) \times a(Oak) + \sum S(F) \times a(Fa) + N(P) \times a(P)$
$= 147.25 \times 0.03 + 14.64 \times 0.02 + 130.53 \times 0.25 + 79.82 \times 0.60 + 117.33 \times 0.07 + 95.8 \times 0.8 + 1 \times 5$
≈ 155.93 sabin

ABSORPTIVITY OF MATERIALS 材料的吸声率

Materials and Furnishings 材料材质	Absorption Coefficients (a) 吸声系数 (a)		
	125Hz	500Hz	4000Hz
Gypsum board 石膏板	0.10	0.05	0.03
Glass/heavy plate 玻璃/厚	0.06	0.03	0.02
Concrete Block/coarse 混凝土/粗糙	0.36	0.31	0.25
Carpet (1/8") 地毯 (3mm)	0.15	0.15	0.55
Oak 橡木	0.58	0.07	0.07
Fabric 纤维织品	0.07	0.49	0.80
Adult (Per person) 成年人（每人）	2.50	4.20	5.00

Sources:
http://www.sae.edu/reference_material/pages/Coefficient%20Chart.htm
http://www.acousticalsurfaces.com/acoustic_IOI/101_13.htm
http://gsd.ime.usp.br/~yiIi/RT60/absocoef.html
http://www.acousticalsurfaces.com/acoustic_IOI/101_13.htm

BEDROOM PLAN 卧室平面图

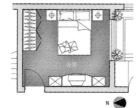

STEP4: CONCLUSION
步骤4: 总结
经过多次尝试和反复计算，卧室内125Hz、500Hz和4000Hz下的混响时间均在预估范围之内。本设计最终选定的材料保证了卧室的声环境质量。使得居民在卧室内可以享受舒适的睡眠和活动质量。

比较：混响时间均在预期范围之内
125Hz: Tr(min)=0.42s <0.45s <Tr(max)=0.70s
500Hz: Tr(min)=0.41s <0.48s <Tr(max)=0.53s
4000Hz: Tr(min)=0.36s <0.43s <Tr(max)=0.44s

STEP3: CALCULATE THE REVERBERATION OF THE BEDROOM ROOM
步骤3: 计算卧室在不同频率下的混响时间
$Tr = (0.05 \times V)/A$ $V=1352.55$ cubic feet
Tr (125Hz) (0.05×1352.55)/151.82≈0.45s
Tr (500Hz) (0.05×1352.55)/139.55≈0.48s
Tr (4000Hz) (0.05×1352.55)/155.93≈0.43s

综合奖・优秀奖
General Prize Awarded・
Honorable Mention Prize

注 册 号：4054
项目名称：阳光加法（黄石）
　　　　　Sunshine Plus (Huangshi)
作　　者：刘宇、王玉婕、桑雨岑、
　　　　　高小燕
参赛单位：重庆大学

阳光加法 Sunshine Plus
The Industrialization Design of Rural Housing in Huangshi Residential Park
1 Background Analysis

District Analysis

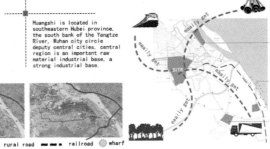

Huangshi is located in southeastern Hubei province, the south bank of the Yangtze River, Wuhan city circle deputy central cities, central region is an important raw material industrial base, a strong industrial base.

— national road　— provincial road　— rural road　--- railroad　● wharf

Climate Analysis

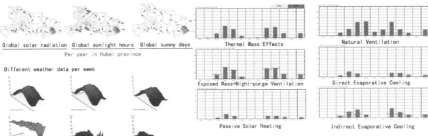

Huangshi has four distinct seasons and moderately humid climate. It is abundant in rainfall concentrating and features a subtropical monsoon climate.

CONCLUSION —— according to the analysis, the design should pay more attention to thermal mass effects, exposed mass+night-purge ventilation and natural ventilation, which have obvious effect on green design.

设计说明：
黄石为典型的夏热冬冷地区。设计提取了当地传统民居的空间特色元素——天井，将其作为设计的核心。绿色方面，采用屋面和地面架空通风带走热量和湿气，并设置风帽加强夜间通风。主动技术采用太阳能集热与光伏发电，达到生态效应最大化，同时结合规划对雨水收集系统进行设计；产业化方面，预制的"天井"构件组装后可根据安放位置不同分别作为天井和阳光间，而拔风西晒墙构件可任意组装满足大小墙体，产业化系列构件既可用于新房建设也可用于旧房改造，具有很强的适应性。

Design Description:
Huangshi is located in typical hot summer and cold winter area. We select courtyard-patio as the core of the design, which is extracted from local traditional houses. In order to ventilate and dehumidify the air from morning till night, multiple measures are taken such as overhead double-top roof and wind caps. Solar collector and photovoltaic system are used to maximize ecological benefits. What's more, rainwater collection system based on village planning is also designed to lead villagers to save energy.

Allowing for industrialization, we designed a series of components. One kind based on courtyard-patio can also become a solar house and act as a buffer layer when installed on building facades. Another kind of components acts as a screen and shading device on west side and is beneficial to ventilation at night in the meantime. Both components can be used not only in the construction of new house, but also in reconstructions in villages.

PERSPECTIVE IN WINTER

阳光加法 Sunshine Plus | 2 Plan Layout
The Industrialization Design of Rural Housing in Huangshi Residential Park

Contextual Analysis

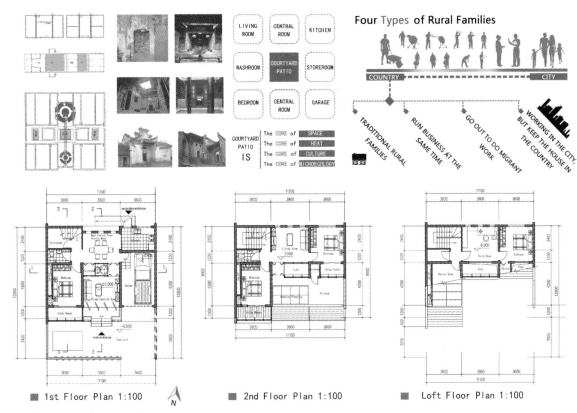

- 1st Floor Plan 1:100
- 2nd Floor Plan 1:100
- Loft Floor Plan 1:100

Adaptability Analysis

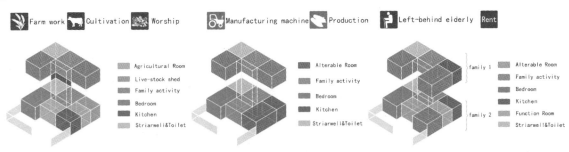

For Agricultural | For Individual Household | For Left-behind

West Elevation 1:100 | South Elevation 1:100 | East Elevation 1:100

阳光加法 Sunshine Plus | 3 Characteristic Techniques
The Industrialization Design of Rural Housing in Huangshi Residential Park

PERSPECTIVE IN SUMMER

Distinctive Tactics Analysis & Temperature Distribution Diagram

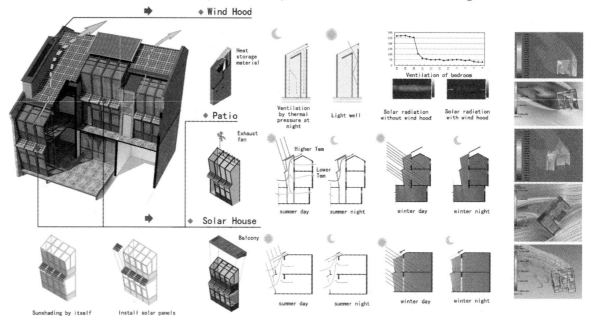

Seasonal Strategies

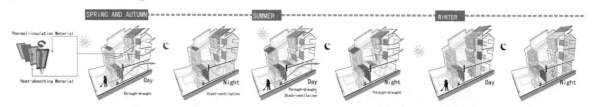

阳光加法 Sunshine Plus | 4 Passive Techniques

The Industrialization Design of Rural Housing in Huangshi Residential Park

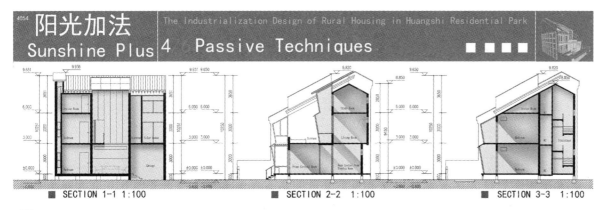

■ SECTION 1-1 1:100   ■ SECTION 2-2 1:100   ■ SECTION 3-3 1:100

Passive Techniques Analysis

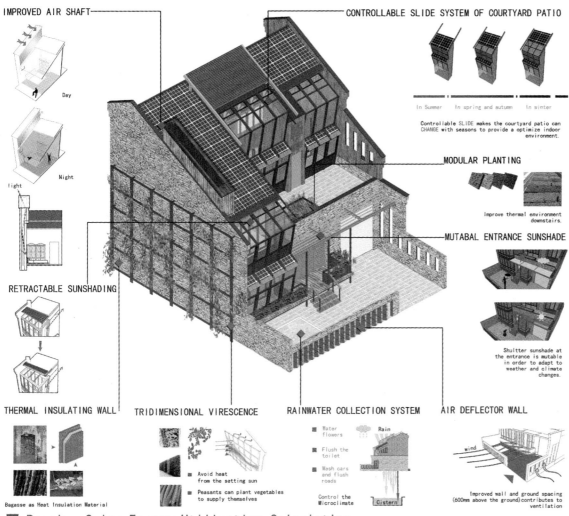

Passive Solar Energy Utilization Calculation

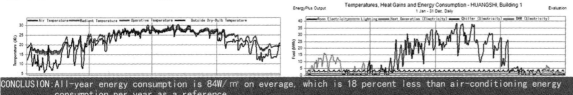

CONCLUSION: All-year energy consumption is 84W/m² on everage, which is 18 percent less than air-conditioning energy consumption per year as a reference.

Sunshine Plus 阳光加法 — 5 Active Techniques

The Industrialization Design of Rural Housing in Huangshi Residential Park

- Solar Water Heating System
- Photovoltaic System

Solar Water Heating System Analysis

- Collector Area
- Solar Paths at Huangshi
- Calculation of System

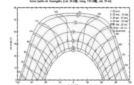

Family Member	Water Demand (L/Day per person)	Solar Utilization Efficiency	Collector Area (m²)	Storage Tank (m³)
6 People	60	50%	$A_{jc} = \dfrac{q_r m C_\rho (t_e - t_L) f}{J_T \eta_{cd} (1 - \eta_L)}$ = 7.2	0.4

According to the solar path at Huangshi and rural background, we designed Solar Energy Utilization plan, assuming that the family is made up of an old couple, a young couple and two kids based on the research of village family constitute.

Photovoltaic System Analysis

- Solar Cell Area
- The Menstrual Solar Power Generation/Solar Loss
- Calculation of Payback Time

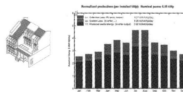

Photovoltaic System	Cost (Yuan/kW)	Roof Area (m²)	Expected Annual Generation Capacity (kWh)	Electricity Saving (400kWh per month) (Yuan)	Subsidy Saving (0.52Yuan per kWh)
66kW	10000	60	5737	2736	487
Net Income per year (Yuan)	Payback Time (Year)	Design Working life (Year)	Earnings in 25Years (Yuan)	ROI	ROI per year
3223	20	25	80100	133%	5.3%

The design working life of system is 25 years. With the government subsidies (0.52Yuan per kWh), It's able to pay back in 20 years.

Industrialization of Patio and Wind Hood

- Patio/Solar House
- Wind Hood

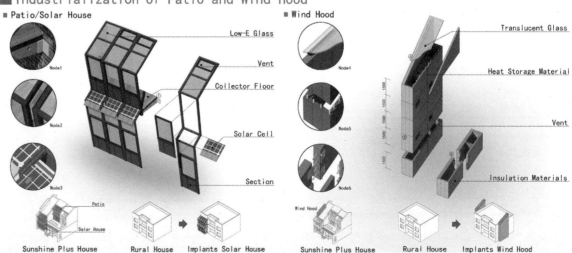

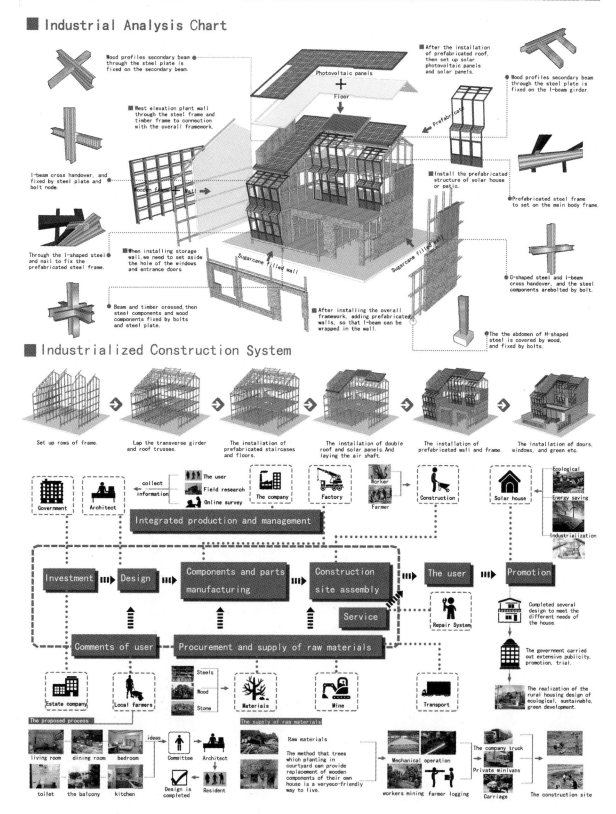

综合奖·优秀奖
General Prize Awarded · Honorable Mention Prize

注 册 号：4142
项目名称："幸·盒·福"（黄石）
"Love·Box·Home"
(Huangshi)
作　　者：刘大用、马镇宇、王晨杨、吴昌亮、孙 杰、徐 亮
参赛单位：东南大学、汉能全球光伏应用集团

4142

"幸 · 盒 · 福" "LOVE · BOX · HOME" 02

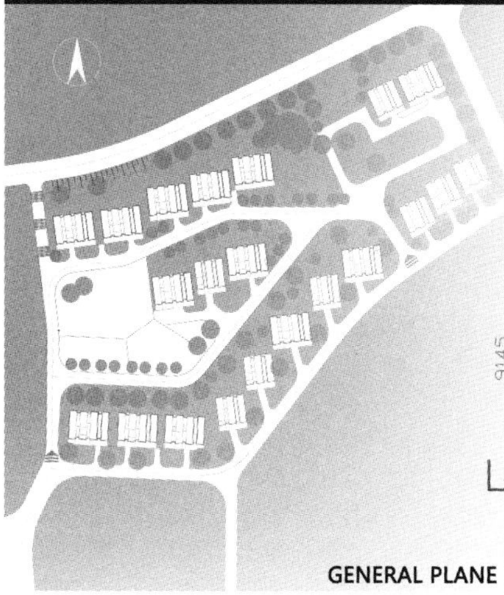

GENERAL PLANE

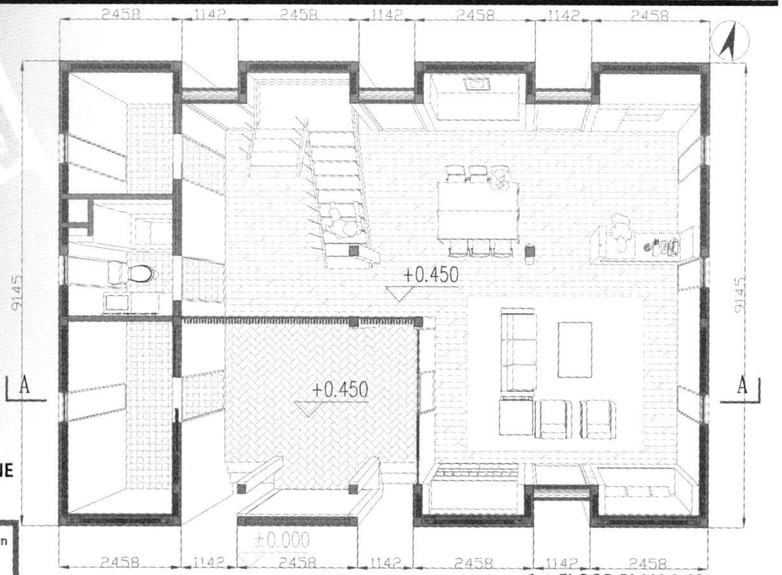
2458
1142
9145
+0.450
A
±0.000

1st FLOOR PLAN 1:40

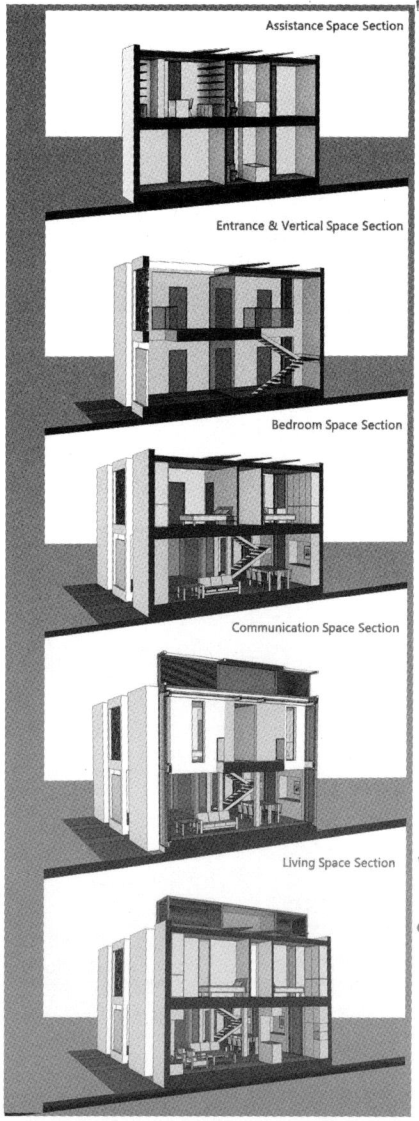
Assistance Space Section
Entrance & Vertical Space Section
Bedroom Space Section
Communication Space Section
Living Space Section

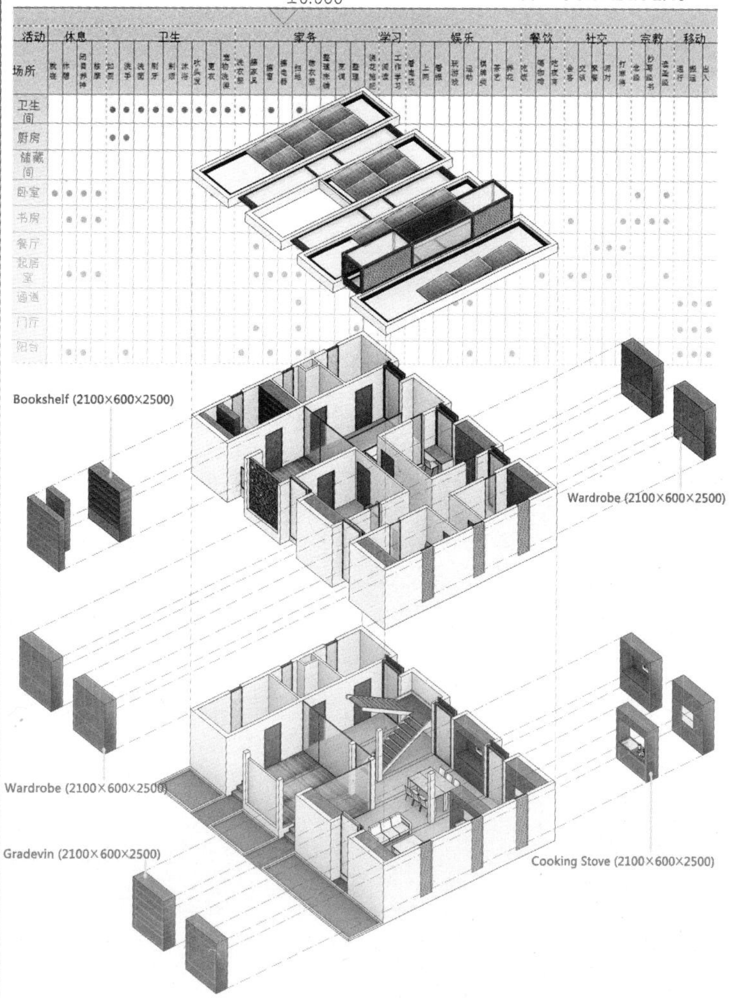
活动
休息
卫生
家务
学习
娱乐
餐饮
社交
宗教
移动
场所
卫生间
厨房
储藏间
卧室
书房
餐厅
起居室
通道
门厅
阳台
Bookshelf (2100×600×2500)
Wardrobe (2100×600×2500)
Wardrobe (2100×600×2500)
Gradevin (2100×600×2500)
Cooking Stove (2100×600×2500)

AXONOMETRIC DRAWING OF FURNITURE

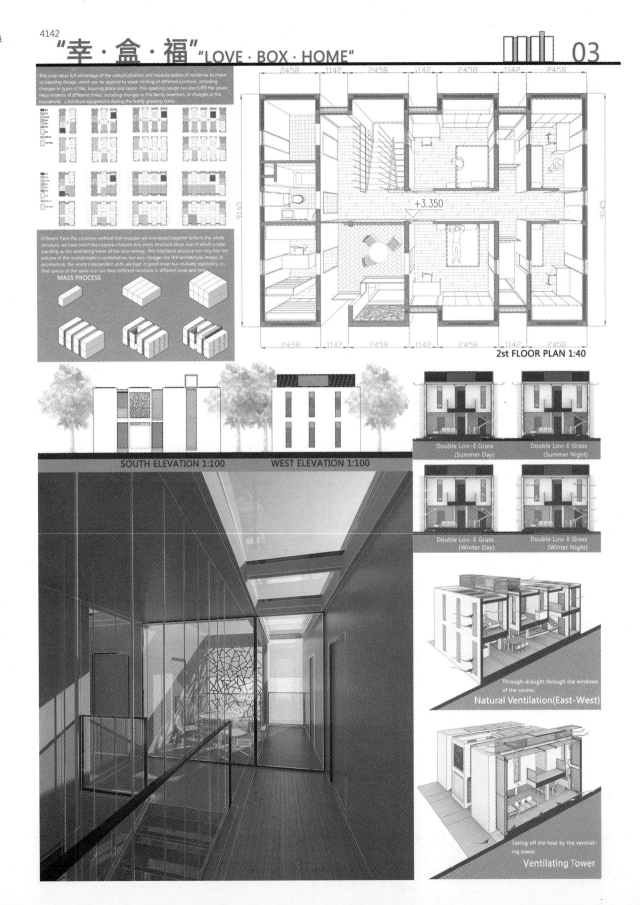

"幸·盒·福" "LOVE·BOX·HOME"

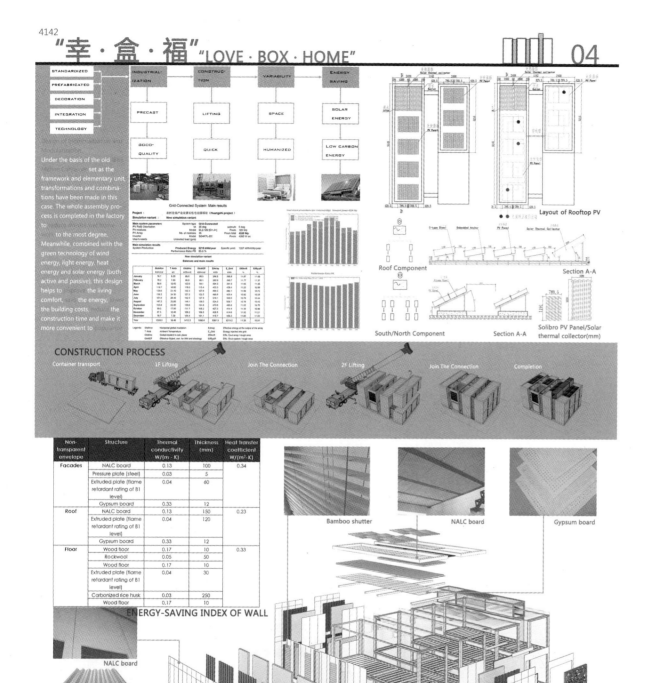

CONSTRUCTION PROCESS

ENERGY-SAVING INDEX OF WALL

AXONOMETRIC DRAWING OF MATERIALS

"幸·盒·福" "LOVE·BOX·HOME"

Window	Type	Structure	Materials of window frame	Heat transfer coefficient
North & South	Low-E hollow glass	6mm Glass+ 12mm Ar 6mm Glass (Low-E-coated)	GRPU	1.5
East & West	Hollow glass	[4mm Glass + 0.2mm Vacuum layer + 4mm Glass] Vacuum glass + 9mm Air+5mm Glass	GRPU	1.8

Window-wall area ratio and Heat transfer coefficient of external windows					
Orientation	Specification	Area (m²)	Window-wall area ratio	Heat transfer coefficient [W/(m²·K)]	Window-wall area ratio limits	K limits
East	Aluminium alloy hollow glass	9.18	0.16	3.6	0.35	4.7
The window-wall area ratio and Heat transfer coefficient meets the requirements in *Design standard for energy efficiency of residential buildings in hot summer and cold winter zone* (JGJ134-2010)						
South	Thermal break aluminum low-E	28.67	0.35	2.3	0.45	3.2
The window-wall area ratio and Heat transfer coefficient meets the requirements in *Design standard for energy efficiency of residential buildings in hot summer and cold winter zone* (JGJ134-2010)						
West	Aluminium alloy hollow glass	9.18	0.16	3.6	0.35	4.7
The window-wall area ratio and Heat transfer coefficient meets the requirements in *Design standard for energy efficiency of residential buildings in hot summer and cold winter zone* (JGJ134-2010)						
North	Thermal break aluminum low-E	18.6	0.23	2.3	0.4	4
The window-wall area ratio and Heat transfer coefficient meets the requirements in *Design standard for energy efficiency of residential buildings in hot summer and cold winter zone* (JGJ134-2010)						

According to the daylight climate zoning regulations in Standard for daylighting design of buildings (GB 50033), Huangshi belongs to the fourth daylight climate zone.
Each room of our model meets the following requirements of daylighting in Standard for daylighting design of buildings (GB 50033):
Bedrooms, living rooms and kitchens in residential buildings should get direct lighting.
Natural illumination in bedrooms, living rooms should be controlled higher than 300lux.
Natural illumination in kitchens should be controlled higher than 300lux.
Natural illumination in bathroom, hallway, dining room, stairwell should be controlled higher than 150lux.

Air permeability performance of exterior windows and transparent curtain walls		
Storey	Air permeability performance rating	Limits
1	4	No less than 4
2	4	No less than 4
Curtain walls	4	No less than 3

meets the requirements in *Graduation and test method for air permeability performance of windows* (GB 7017-2002) and *Graduation of physical performances for building curtain walls* (GB/T 15225)

Visible light transmittance of exterior windows			
Orientation	Window-wall area ratio	Visible light transmittance (Tv)	Limits of transmittance	Shading coefficient of exterior window
East	0.16	0.72	0.5	0.72
Visible light transmittance of the windows to the east meets the No. 5.0.3, Tv should be controlled higher than 0.50.				
South	0.35	0.69	0.5	0.61
Visible light transmittance of the windows to the south meets the No. 5.0.3, Tv should be controlled higher than 0.50.				
West	0.16	0.72	0.5	0.72
Visible light transmittance of the windows to the west meets the No. 5.0.3, Tv should be controlled higher than 0.50.				
North	0.23	0.69	0.5	0.61
Visible light transmittance of the windows to the north meets the No. 5.0.3, Tv should be controlled higher than 0.50.				

SHADOW ANALYSIS WIND ANALYSIS ENERGY-SAVING INDEX OF WINDOW

A-A SECTION 1:30

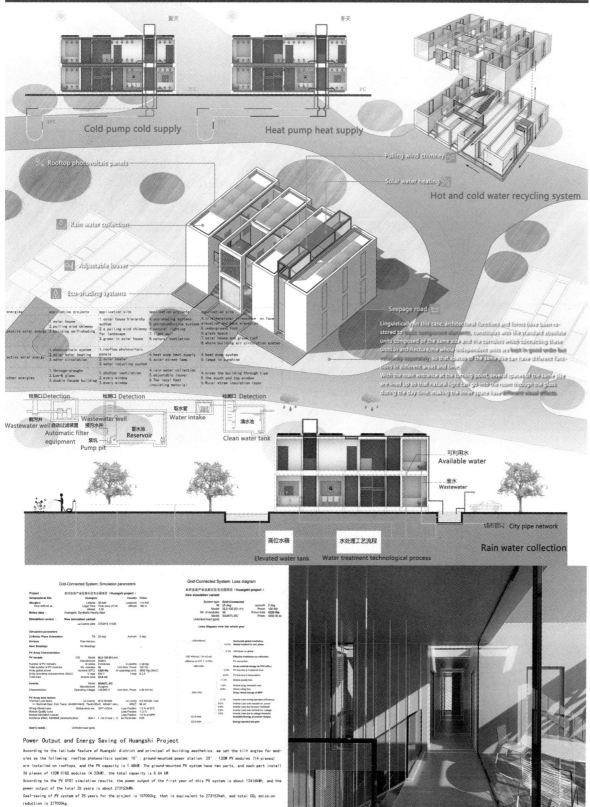

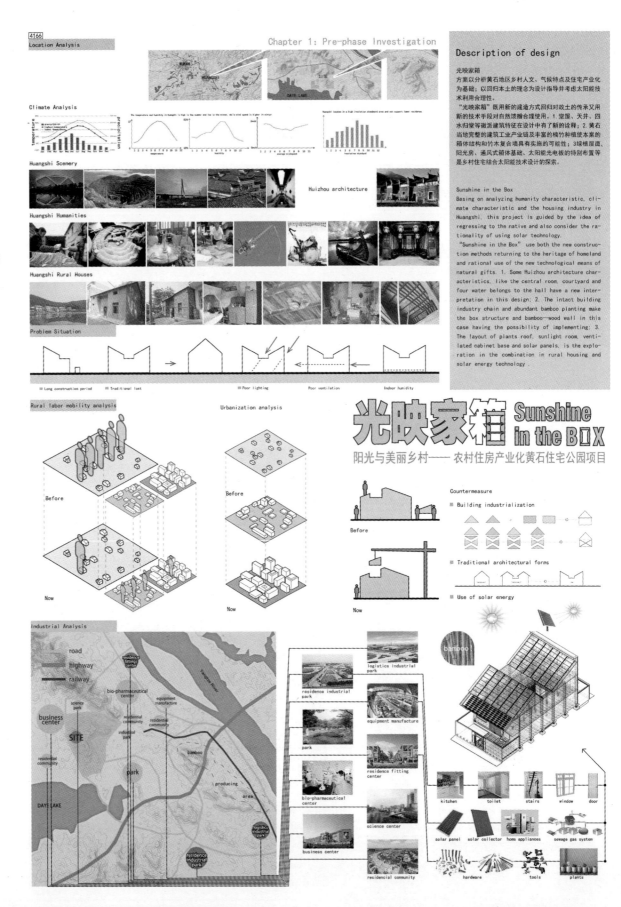

Box concept origin

Local traditional houses plan

Modular system basis

Modular system
1M=100mm
3M 6M 12M 15M 30M 60M
1/10M 1/5M 1/2M

Dinning room   3600mm X 2400mm
Kitchen   3600mm X 2400mm
                3600mm X 1200mm
Wash room   3600mm X 2400mm
                2400mm X 1200mm
Patio   3600mm X 2400mm

Truck carriage (10t - 15t)
2400mm X (8100mm - 8600mm)

DELTA Amorphous silicon thin film solar cells
1245mm X 635mm X 7.5mm

Plate solar collector Sanggao
2000mm X 1000mm X 90mm

Chapter 2: Concept of The Box

Isometric view of Cabinet structure
a Box structure
b Wall structure
c Wall decomposition structure
d Partition structure
e Partition wall decomposition structure
f Wall
g Wall decomposition structure
h Ceiling structure
i Floor structure
j Construction parts — Aquarium
k Construction parts — Stairs
l Construction parts — Washroom
m Construction parts — Kitchen

Deconstruction plan

Restructured plan

Regenerate plan

Box exploded isometric view

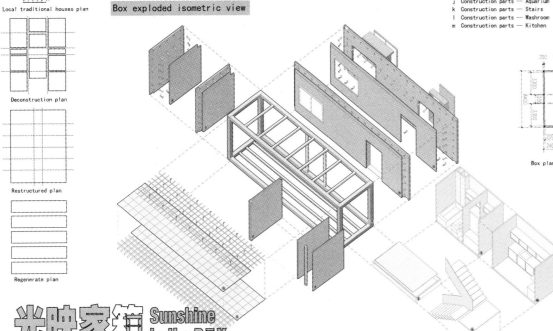

Box plane size

光映家箱 Sunshine in the BOX
阳光与美丽乡村——农村住房产业化黄石住宅公园项目

Assembled

 Basic box A
 Basic box B
 Basic box C
Basic box D

Construction parts

Cabinet interior decoration

 Basic box E
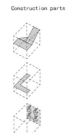 Basic box F
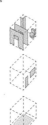 Roof box
Sunroom box

3 boxes
1B+1C+1F

4 boxes
2B+1E+1F

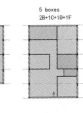
5 boxes
2B+1C+1D+1F

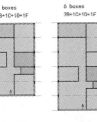
6 boxes
3B+1C+1D+1F

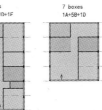
7 boxes
1A+5B+1D

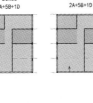
8 boxes
2A+5B+1D

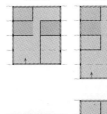
9 boxes
1A+4B+2C+1D+1F

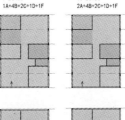
10 boxes
2A+4B+2C+1D+1F

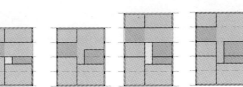

→ Entrance

 Structure
Stair
Wash room

Living room / Principal room
Patio
Kitchen
Bedroom

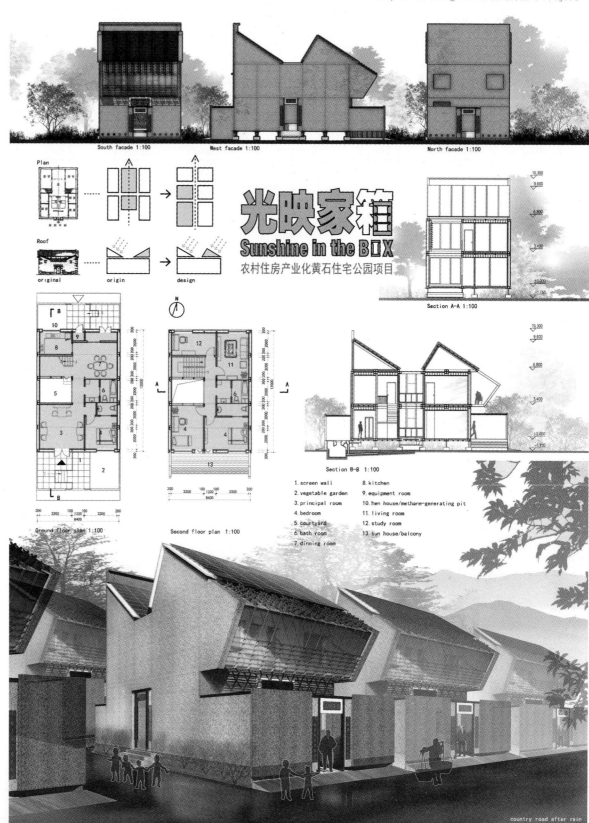

Utilize of Bamboo

- Acquire material conveniently, rich yield
- Low price, environmental protection
- Heat preservation and heat insulation safe fireproofing

Chapter 4: Passive Solar Tactics

光映家箱
Sunshine in the BOX
农村住房产业化黄石住宅公园项目

Technical Analysis

| Summer Day | Summer Night | Heat Insulating in Summer | Sunshade in Summer |
| Winter Day | Winter Night | Heat Insulating in Winter | Sunshade in Winter |

Ecotect Analysis

Natural Light Illumination   Range: 0.00~0.80m/s

Air Flow Rate   Range: 0.00~0.80m/s

Air Flow Rate   Range: 0.0~3.0m/s

Air Flow Rate   Range: 0.0~5.0m/s

Hand Crafted Model

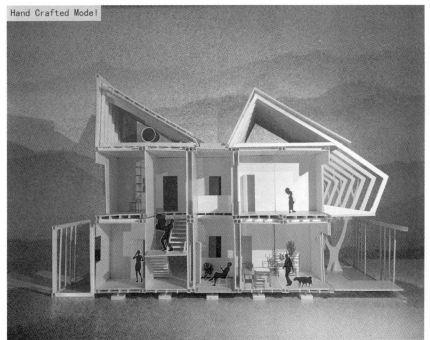

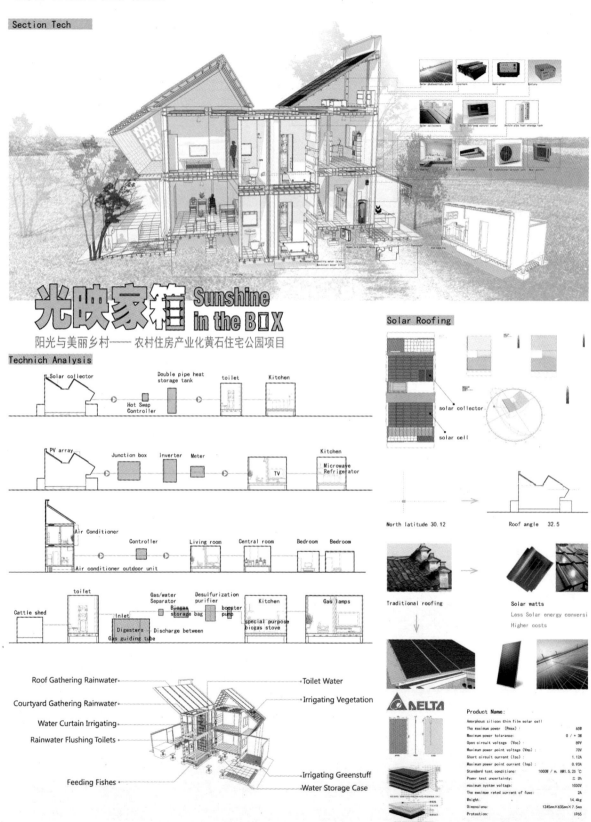

Chapter 6: Possibility of Industrialization Construction

On-site Assembly Process

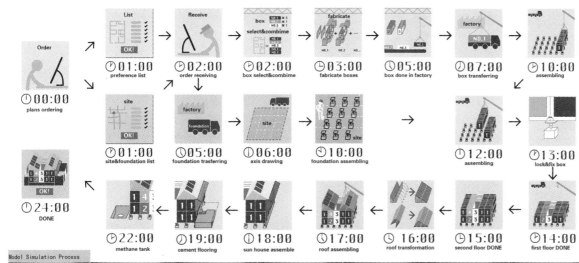

Model Simulation Process

光映家箱 Sunshine in the BOX

阳光与美丽乡村——农村住房产业化黄石住宅公园项目

Statistics

Building Foot Print: 151.2㎡	Floor Area Ratio: 1.37
Total Area: 207.5㎡	Building Height: 10.8m
Courtyard Area: 49.2㎡	Number of Floor: 2

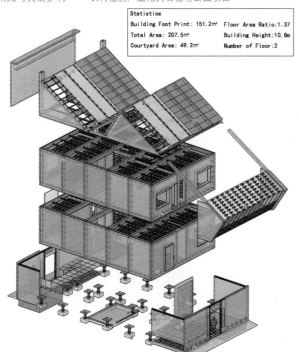

List of Materials

Name	Length	Number
Whole boxes		
200mm×200mm Square Bar	8000mm	57
200mm×200mm Square Bar	2400mm	112
100mm×100mm Universal beam	3000mm	24
100mm×100mm Universal beam	2000mm	60
100mm×100mm Universal beam	8000mm	30
External wall components, Angle steel and fastener		1176
Bamboo and wood composite wall	8400mm×3400mm	23
Bamboo and wood composite wall	2400mm×3000mm	16
Ceiling rod with 200 height, deformed steel bar, angle steel;		32 (set)
Lightweight steel keel with 25mm thickness	8400mm×3000mm	8.5
Lightweight steel keel with 100mm thickness	8400mm×3000mm	8.5
Ceiling with plasterboard	8400mm×3000mm	8.5
Bamboo and wood composite floor	8400mm×3400mm	8.5
Polyethylene insulation board	8400mm×3400mm	4.5
Two layers of macromolecule waterproof material	8400mm×3400mm	5
Small gray tiles hung by machine	8400mm×3400mm	5
Cement brick outside	8400mm×3000mm	2.5
Precast concrete base	600mm×600mm×300mm	24
Adjustable spiral base platform		24
Window	1800mm×1500mm	5
Door	1200mm×2200mm	1
Door	1200mm×900mm	7
Door	1200mm×750mm	6
Window	8000mm×2000mm	1
Planting	8400mm×6800mm	1
Planting	3000mm×6800mm	1
Three stacked amorphous silicon battery pack (DELTA)	1245mm×635mm×7.5mm	40
Solar thermal pannel	1000mm×2000mm	2

Detail Structure

Roof drainage structure | Wall hanging structure | Box connecting structure 2
Roof structure 1 | Cabinet floor structure | Patio roof structure
Roof structure 2 | Vent structure | Wall hanging structure
planted roof structure | Panel mounting structure | Wall hanging structure
Cabinet ceiling construction | Box connecting structure 1 | Basic structure

2015台达杯国际太阳能建筑设计竞赛获奖作品集

综合奖·优秀奖
General Prize Awarded · Honorable Mention Prize

注 册 号：4241
项目名称：生·升不息（黄石）
　　　　　Ecology Evolution (Huangshi)
作　　者：应振国、曹世彪、王晓楠、
　　　　　诸葛涌涛
参赛单位：天津大学、无唯工作室

聚落是建筑、环境和空间等物质系统，同时也是文化系统、社会系统等非物质系统的综合。

中国农村不断发展人口比例达到50%，拆旧房建新房成为农村盛行的一种现象，从而保证了聚落形态长久不变。"**再现**"两湖聚落或当地原有村落的传统意象，避免机械化布局，唤起居民的情感记忆，"并非在原地拆掉重建"，而是**有机生长**，保证聚落的延续与发展。

1. PROGRAM

　　Settlement is a synthesis of material system such as architecture, environment and space, and immaterial system such as culture system and society system.
　　The village of China is in a constant development, and the proportion of rural population is 50%. Pulling down old houses and building new houses become a prevalent phenomenon, and that guarantee a secular stability of settlement pattern. "**Reappear**" the settlement pattern of hubei province or old pattern, avoid mechanized plan, recall the emotional memory of residents. Not build new house, but **organically develop**, to ensure the continuation and development of settlements.

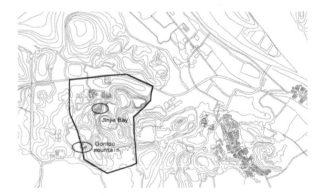

Jinjia Bay　　　Goutou mountain
old settlement

current rural landscape

organic renewal of old settlement

生·升不息
Ecology Evolution

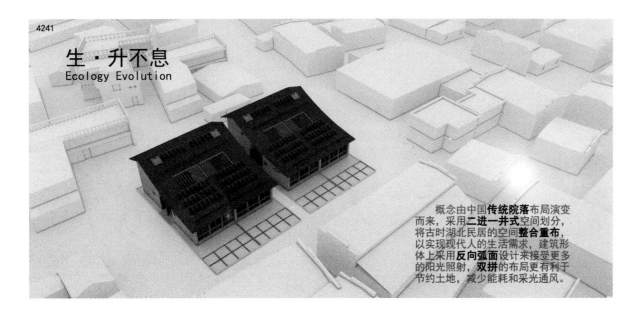

概念由中国**传统院落**布局演变而来，采用**二进一井式**空间划分，将古时湖北民居的空间**整合重布**，以实现现代人的生活需求，建筑形体上采用**反向弧面**设计来接受更多的阳光照射，**双拼**的布局更有利于节约土地，减少能耗和采光通风。

2. CONCEPTION

The concept is evolved from the traditional **Chinese cuontyard**, which takes use of the form of **two rooms with one countyard** and it makes the ancient Hubei folk houses **rearrange** to realise the need of the modern people's life. The form of the architect takes the shape of **reverse cambered surface** to accept more sunshine. Additionally, The **binary house** design can save much more land, reduce the use of the energy and it is more advantageous for the ventilation and lighting agree.

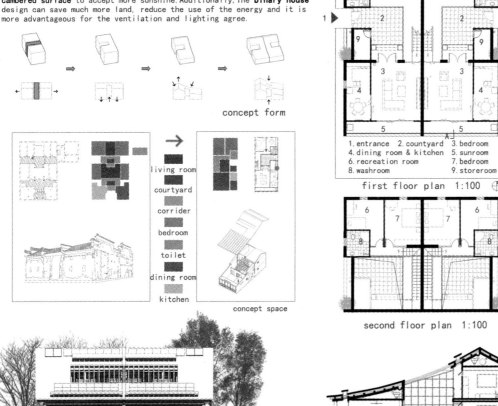

1. entrance 2. countyard 3. bedroom
4. dining room & kitchen 5. sunroom
6. recreation room 7. bedroom
8. washroom 9. storeroom

first floor plan 1:100

second floor plan 1:100

south elevation 1:100

section A-A 1:100

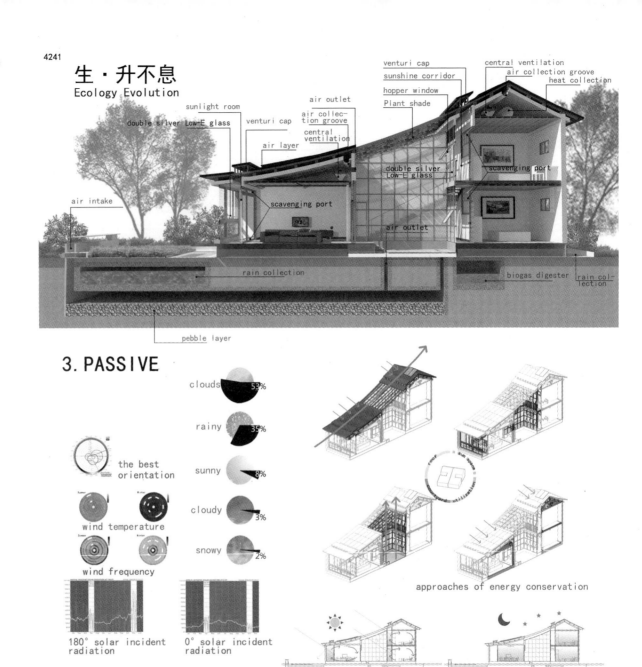

生·升不息
Ecology Evolution

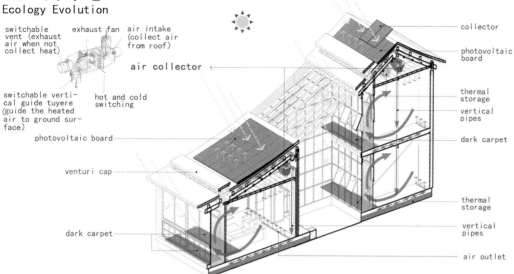

4. COMPREHENSIVE

Principle of work: In the day of the winter, the solar irradiation heats the air, and then it is taken to the room and a part of them is stored in the thermal storage material floor and tank,
In the night of the winter, that quantity of heat is released to heat the room.
In the day of the summer, open the ventilation system and accelerates the heat dissipation of the roof, and store some irradiation to heat water.
In the night of the summer, it can take in cool wind to bring pleasantly cool.
In a nutshell, this solar system can realize multifunction.

daylight in winter

In winter, the roof interlayer collect heat, guide the warm air to the floor surface by the ventilating device, and heat the room.

daylight in summer

In the daylight of summer, the roof interlayer exhale the hot air to outside by ventilating device to realize heat insulation.

night in summer

In the night of summer, the air in outside is cool, the ventilating device can take in the cool air and exhale the hot air inside.

Accord by the analysis of shadow range, the south-north distance of two "townhouse" is ultimately intended to be 12.3m. The result is, in the spring, summer and autumn, the south courtyard can get enough solar irradiation to guarantee the growth of the vegetables. and in winter, there will not too much shade at south window.

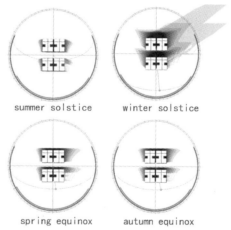

summer solstice | winter solstice
spring equinox | autumn equinox

Building superficial area & volume (of two families)

Atotal surface area=729.42m²
Ayard surface area=104.6m²
Asurface area=624.82m²
A/V Ratio=624.82/1189=0.526 < 0.55

Vtotal=1906.5m³
Vyard=717.5m³
Varch=1189m³

superficial area of building

Ssouth wall area=57.4m²
Seast wall area=116.25m²
Snorth wall area=114.8m²
Swest wall area=116.25m²
Sroof=324.72m²

winter | summer

direct solar irradiation on the roof

The quantity of electricity generated by the photovoltaic system can meet the fundamental demand of the residents.

生·升不息
Ecology Evolution

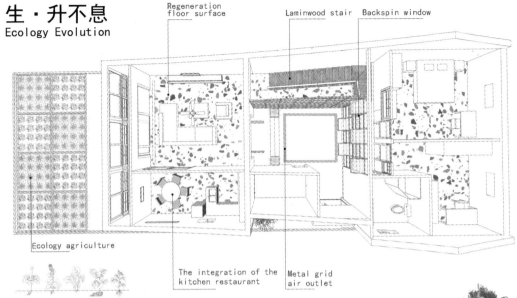

east elevation 1:100

Regeneration wall surface

Regeneration floor surface

5. MATERIAL

On the aspect of the choose of materials about our design, we take use of the local old buildings ' materials and we recycle as well as reuse the local materials and process them in the factory. By using these local materials, we can reduce more energy and save much more time and money. What's more, the way that we take use of the local houses' materials is more environment-friendly and it is also lower carbon in the procession of collection, production and the transportation.

The wall & floor materal

heat transfer coefficient in the maintenance structure roof: K=0.236

area	hot summer and cold winter area standards	Heat transfer coefficient of building maintenance structure
heat transfer coefficient in the maintenance structure roof	≤1.0	0.236

The table shows that conform to the standard.

exterior wall outside heat transfer coefficient in the maintenance structure: K=0.359

area	hot summer and cold winter area standards	Heat transfer coefficient of building maintenance structure
exterior wall outside heat transfer coefficient in the maintenance structure	≤0.8	0.359

The table shows that conform to the standard.

生·升不息
Ecology Evolution

节水80% Save of water
节树20% Save tree
节能70% Save of energy
节地20% Save of the land
节时70% Save time
低碳环保 Low-carbom environment

design

experiment

producting

physical distribution

assembling in the site

6. INDUSTRIALIZATION

In order to realize rural housing industrialization in the hot-summer and cold-winter zone, we use a structure of wood and steel. The use of steel is convenient to prefabricate and bear load, and the use of wood can conserve energy, and it is low-cabin, ecology and comfortable in visual sense.

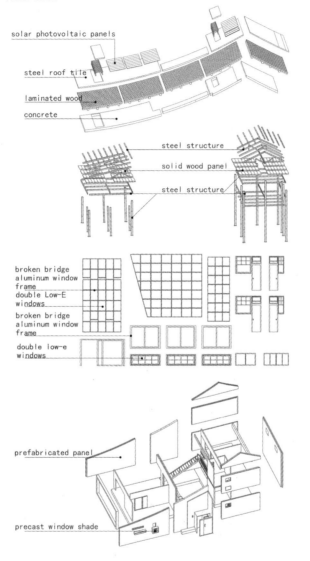

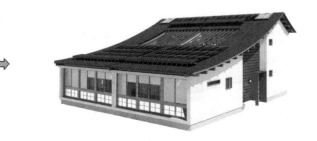

综合奖 · 优秀奖
General Prize Awarded · Honorable Mention Prize

注 册 号：4266
项目名称：材料之"煜"——乡村太阳能生态住宅设计（黄石）
The Glorious Sunshine of Materials—Rural Solar Ecological Residential Design (Huangshi)
作　　者：刘罡、刘紫伊、梁鑫、邵继中
参赛单位：南京工业大学

材料之"煜"
Rural Solar Ecological Residential Design
——乡村太阳能生态住宅设计

NO. 4266

Concept

本设计从材料出发，采用了自然界中极其容易获得的元素——黏土、植物和阳光建成，紧紧地与大地融为一体。而它又是一所不简单的生态住宅，它收获着阳光、水、风等自然资源，转化成人们需要的各种能源，使家庭生活更自然健康。

This design based on materials. It is a simple house which is made from the most foundamental elements in nature—soil, plant and light. It seems like the house was born on the ground. The house is not a simple rural house. It absorbs sunlight, water and wind from nature and converts the solar energy into other kinds of energy we need by using corresponding techniques.

Masterplan

Question
1. What material should we choose which causes less environmental damage?
2. How to combine the solar energy with the design?

Analysis
The local buildings in Huangshi face varieties of problems, such as winter heating, cooling in summer and humid air.

So we should choose materials which adapt to circumstances, and take advantages of solar energy.

Answer
Clay is the ideal choice.

Modern buildings rely heavily on industrial technologies which have to consume large amounts of energy and produce a lot of pollutants. Our design aim to respect the environment, feel the sun, the breeze and the changes of seasons.

concrete　wood　stone　clay　local pictures

Econo-technical Norms

Floor Area:	150m²
Building Area:	219m²
Floor Area Ratio:	1.87
Solar Cell Area:	24m²
Solar Collector Area:	10m²
Heating Area:	149m²
Roof Planting Area:	54m²
Floor Number:	2
Floor Height:	3m

Model photos

材料之"煜"
Physical properties of Materials Analysis
——围护材料物理性能分析

Clay building is built with soil and crops fiber as basic material, which is completely non toxic, and can be recycled. so it has little impact on the environment.

We selected 6 kinds to analyse, hoping to find an ideal building material.

Material selection

	Raw materials	manufacturing process	Advantages	Disadvantages
麦秸泥土 cob	Sand, Marls, Straw, Wheat, water	Mix without any equipment.	Good heat insulation. Keep the humidity indoor. Produce on the spot. Convenient transport.	dampness
夯筑泥土 Compacted soil	Soil, Cement as stabilize	Set Templates, pour the soil and compact, and layer.	Good heat insulation. Good Waterproof performance. Durability is good.	Strength, dampness
袋装泥土 Bagged soil	Textile bags or Plastic bags, soil	Put the soil in the textile bags or plastic bags.	Low cost. Free form(The remote area, disaster-relief area)	Instability
轮胎夯实泥土 Tire compacting soil	Used Tires, soil	Stagger the tires filled with soil, and layer.	1. Good heat insulation 2. Durability is good 3. Recycle the used tires 4. The strength of the wall is greater than adobe building	fire prevention (daub mud on the surface)
轻质黏土 Light-clay	Crop fiber (Sawdust, coir, Flax, etc.)	Collect the clay, mix with the crop fibers.	1. Adjust the temperature, and humidity 2. Good heat insulation 3. Low cost	Need a long time to dry off.
农作物砌块 Crop fiber block	Crop fiber, polyester fibres, metal wires/wire net	Bind the crop fibers By machine or hand.	1. Good heat insulation 2. Low cost 3. Free form	fire prevention dampness

According to the analysis of environment in Huangshi, the Light Clay is the most suitable material.

Material making

In order to verify our inference and select the most appropriate material, We handmade four clay bricks:

- Clay+Straw
- Clay+Straw+Cottons
- Clay+Straw+Sand grains
- Clay+Sawdust

Besides, a common brick for contrast.

Brick number	Clay	Straw	Sand	Sawdust	Cotton
Brick 1	○	○			
Brick 2	○	○	○		
Brick 3	○			○	
Brick 4	○				

Thermal performance analysis

Time: 12:20–17:50 21st, Dec. 2014

Brick number	Brick 1 (℃)	Brick 2 (℃)	Brick 3 (℃)	Brick 4 (℃)	Common brick(℃)
12:20	7.8	8.9	8	9	9.3
14:50	13.5	12.1	13.8	15.1	15.4
17:20	8.6	8.7	9.2	9.3	8.5
Heat storage Δ (12:20~14:50)	5.7	3.2	5.8	6.1	6.1
heat radiation Δ (14:50~17:20)	4.9	3.4	4.6	5.8	6.9

The minimum thermal resistance of building insulation design: $R_{o \cdot min} = \frac{(t_i - t_e)n}{[\Delta t]} R_i$

Relationship between materials and thermal performance

→ 12:20 ■ 14:50 ▲ 17:20

Results: In consideration of the hot properties and other composite factors of all the bricks in the experiment, the brick which is made of clay, straw and cotton, has little influence of solar radiation and the best thermal performance, so we chose it as the ideal material.

Mechanical analysis

Materials:
Brick 1: Clay Brick 2: Clay+Straw
Brick 3: Clay+Sand Brick 4: Common brick

Method: The direct shear test

Brick1 ◆ Brick4 ● Brick2 ■ Brick3 ▲

Results: brick1 has the best shear strength. when there is larger crop fiber content in the brick, the shear strength of bricks become smaller. And, the brick contents straws has larger shear strength than the sand.

Sticky rice juice

Removes the impurities

Clay

Add sticky rice juice

Straws

Stirring

Sawdust

Adding clay in the mode

Sand grains

Cottons

Adding cottons

Carbon fiber cloth

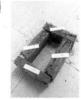
Adding carbon fiber cloth

Natural air-dry

Putting the hay on the surface

Process of making Light Clay

Materials
Process
Measurement

Temperature measurement

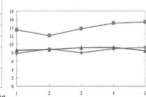
Nomber the bricks

The direct shear test

Prefabricated Structures Analysis
材料之"煜" —— 结构与构件工业预制化分析 3

NO. 4266

Wooden Skeleton Structure

Labels:
- The wooden keel of sloping roof
- Diagonal roof support
- Wood floor
- Clay Wall
- Rainwater recycling
- Roof cavity
- Window installation
- Secondary beam
- Concrete foundation

Concept

1. Hope

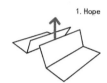
2. Programme Mass

3. Solar Energy

4. Wind

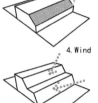
5. Water Folowing

Wooden Structure Analysis

① 240 Clay brick
② 180 Wood keel filled clay layers.
③ Cracking mortar composite alkali fiberglass mesh Plaid
④ Elastic primer, flexible water resistant putty
⑤ Exterior paint

① Wood floor
② Plastic film
③ Cotton insulation
④ Waterproofing membrane
⑤ Concrete pads
⑥ Strip foundation

① Planted roof
② Waterproofing membrane
③ Wood keel
④ Roof cavity
⑤ Cement board

Local materials and prefabrication

Concrete foundation

Connection between precast

Integrated from local timber, large-scale production of prefabrication.

The prefabricated girders beams and local production of steel plants were connected to the frame assembly.

Local waste wood can be treated to make some non-renewable materials used for load-bearing wood components.

Large quarry (Steel quarrying plant) can be used to produce steel link members.

Small quarry, stone clay foot of wall can be used as fill material department.

The extensive use of local coal ash bricks.

Local houses being built in the form.

1	Living Room
2	Kitchen
3	Dining Room
4	Storage Room
5	Bed Room
6	Toilet
7	Entertainment
8	Solar House

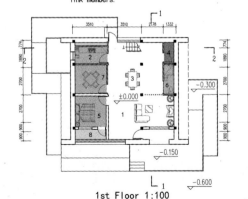
1st Floor 1:100

2nd Floor 1:100

材料之"煜"
Residential Environment Simulation Analysis
——建筑环境模拟分析

Geographic Position

Yellowstone Park residential project is located in Huangshi city, Hubei province, the south bank of the middle reaches of the Yangtze River.

The annual average temperature is 16.4°C, extreme maximum summer temperature 39.4°C, winter extreme minimum temperature -18.1°C, the hottest month average of 29.2°C, the coldest month average of 3.9°C.

The annual average rainfall of 1382.6mm, the average annual rainfall in about 132 days, the daily maximum rainfall of 360.4mm.

Ventilation Parameters

参数	夏季	冬季
空气调节日平均温度（℃）	35.3	-2.4
空气调节计算日较差温度（℃）	29.4	
空气调节计算日温度（℃）	32.2	
通风计算日温度（℃）	32.0	0.1
空气调节计算日相对湿度（%）	63	72
平均风速（m/s）	2.0	2.6
风向	东南风 SE	东北偏北 NNE

Basic Meteorological Data

Meteorological parameters:
latitude 29°30'~30°15', longitude 114°31'~115°30',
166 meters above sea level measurement.

Summer Wind analysis

The design focuses on the need to solve the problem of ventilation and cooling in summer.

After calculation, the wind flow in the blue region of the largest and strongest wind in the south east of 23°.

Annual daylighting analysis

After calculation, it can be seen from the figure: from January to December of lighting conditions lighting something more ideal surface.

So something light clay wall retaining wall to ensure its maximum utilization of solar energy.

Best thermal radiation

Combined with climatic characteristics.

The design focuses on the need to solve the problem of winter heating collectors.

- Solar radiant heat in summer
- Solar radiant heat in winter
- Annual mean of solar radiant heat
- The median of solar radiant heat

West Elevation 1:100

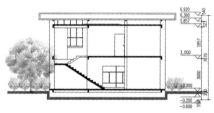

South Elevation 1:100

Section 1-1 1:100

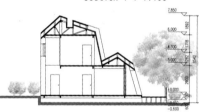

Section 2-2 1:100

材料之"煜" —— 主被动太阳能技术分析
Passive Solar Technical Analysis

Interior Perspective

Summer Ventilation

Summer ventilation mainly rely on the natural convection and hot wind drawing to achieve together.

Summer Heating
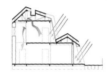
With the summer sunlight into the interior, the temperature rises, taking the heat through the top of the roof away.

Winter Heating

With the sunlight in winter into the room, the heat can flow through the underground cavity and the wall.

Illumination Analysis

Microclimate Analysis

Place a small pool outdoor to create a microclimate system for the house, and improve the local climate.

Roof Pool Heat Storage

Roof pool can provide heat for the house in winter, it can be opened during the day to draw the heat.

Roof Heat Dissipation

Roof pool can be closed at night, the heat stored during the day will loose into the room.

Water Cycling System
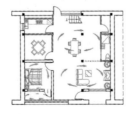
Water circulation system of residential connects the water storage module through the pipelines.

Solar Photovoltaic System

Design of building roof inclination combines of the photovoltaic system, in order to maximize the utilization of solar energy.

Rainwater Collection System
Rainwater is collected and purified through the rainwater collection system and the purification system.

Plane Thermal Cycling
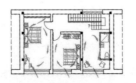

Plane Ventilation Cycling

Materials for Solar Energy Utilization
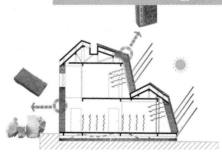
Clay is good for heat preservation and thermal storage.

The wallsare made of clay, covered with roof greening. It makes the building absorb heat as much as possible in winter. The heat is transmitted to inside through the walls. This is good for people to keep warm.

材料之"煜"
Detail Drawing of Structures Analysis ——结构节点大样分析

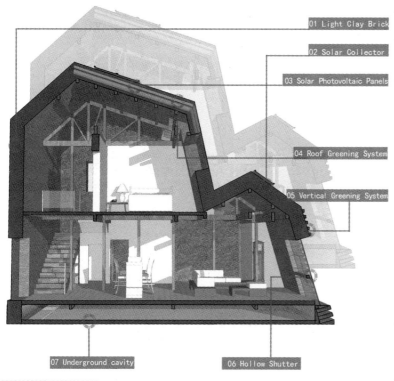

- 01 Light Clay Brick
- 02 Solar Collector
- 03 Solar Photovoltaic Panels
- 04 Roof Greening System
- 05 Vertical Greening System
- 06 Hollow Shutter
- 07 Underground cavity

Delivery Perspective

Light clay is a kind of composite materials, mainly contains the natural crop fiber and clay. Building with clay is a combination of brick masonry, sculptures and cooking and other fun things. Light clay has good heat insulation performance, is an ideal material for passive solar buildings.

- Hay
- Clay
- Cotton
- Carbon fiber cloth
- Clay
- Hay

01 轻质黏土砖 Light Clay Brick

The solar collector has simple structure, reliable operation, low cost, lower heat flux. compared with the vacuum tube, the solar collector has stronger bearing capacity, larger heat absorption area, higher thermal efficiency.

02 太阳能集热器 Solar Collector

Solar photovoltaic power generation system is safe and reliable, no noise, no pollution. It takes the advantages of building roofs, and the construction period is short.

03 太阳能光伏电板 Solar Photovoltaic Panels

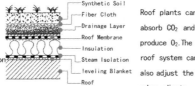

- Synthetic Soil
- Fiber Cloth
- Drainage Layer
- Roof Membrane
- Insulation
- Steam Isolation
- leveling Blanket
- Roof

Roof plants can absorb CO_2 and produce O_2. The roof system can also adjust the urban climate.

04 屋顶绿化系统 Roof Greening System

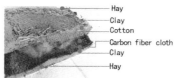

12~17cm
- Microtube irrigation
- Support structure
- Plant box
- Air layer
- Waterspout
- Ground

Vertical green wall can be used for as long as 10 years, is easy to replace.

05 垂直绿化系统 Vertical Greening System

Hollow shutter window can save the space, it can also insulate the thermal and anti noise. No matter it is summer or winter, it can adjust louver angle, so that it greatly reduces the energy of air conditioning.

06 中空百叶玻璃窗 Hollow Shutter Window

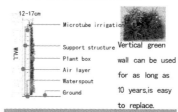

Roof drainage. We set drainage grooves on the roof, then collect rainwater by the sloping roof. The water flows into the water storage module underground.

Water Storage Module Pipelines

07 蓄水模块 Water Storage Module

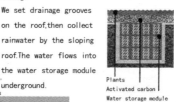

- Plants
- Activated carbon
- Water storage module

Rainwater collection system collects the rainwater through the pipeline, then purify the rainwater into clean water. The water can be used in grass irrigation and the toilet.

Model photos

Gardens in House 01
集院宅

4278

综合奖·优秀奖
General Prize Awarded ·
Honorable Mention Prize

注 册 号：4278
项目名称：集院宅（黄石）
　　　　　Gardens in House (Huangshi)
作　　者：黄绮琪、郑楚烽、陈宗煌、
　　　　　关竣仁、张　铮、赵一平、
　　　　　许蔓灵、莫旎卡、骆武辉
参赛单位：华南理工大学建筑学院青年设
　　　　　计工作室

Background

	Before	Trend
population structure		
agricutural industry		
building industry		
financial structure		

Household poupulation structure will be simplified. Monoculture becomes popular for higher production efficiency while small fields in farmers' houses will be used for personal needs. The traditional construction mode will be replaced by a better one. Farmhouses will become economic factor in rural life.

Design Concept

本方案把产业经营和小型农家种植元素纳入农宅之中，把居家与产业化从建筑本身的形态与建造，衍生到日常的生活与生产当中，大坡屋顶不仅是二层表皮也是纳入光、风、雨、雪的媒介，它模糊了建筑内外的界限，融合了内外空间。

Our design combines the elements of industrial operation and domestic planting with the rural house. Meanwhile, it put industrialization and daily life into the thing which ranges from building itself to living and production. Large sloping roof is not only the dual envelope, but also the carrier containing light, wind, rain and snow, which blurs the boundaries between inside and outside.

Site Analyse

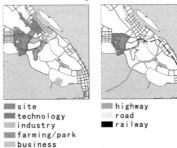

■ site
■ technology industry
■ farming/park
■ business
■ highway
■ road
■ railway

The area surrounding the site has planned the farmland and the business district, which lays the foundation for small-scale domestic planting in the courtyard. Perfect surrounding industrial area and convenient transportation bring advantages for the construction industry.

Industrial Dwelling

1) Household Business Operation

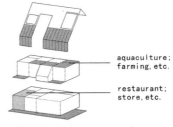

aquaculture; farming, etc.

restaurant; store, etc.

Farmhouses become economic factors in rural life, such as organic rural inns or family-owned store, etc. We also used a small field to do some farming.

2) Building Industry

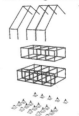

The modular frame, foundation and roof structure

Stacked modular frame forms the basic structure. The house is controlled by the module to be in an orderly state, ranging from the architecture form to construction.

Gardens in House 02
集院宅

Architecture Form

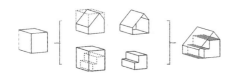

Vague space between interior space and exterior surface is created by reduction of volumes. Integrated with light, wind and water, vague space is set as a three-dimensional garden. A large sloping roof is conducive to collect solar energy and help accumulated snow fall.

About Multipurpose Area

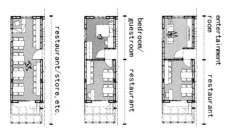

The multipurpose area can be used as the expansion space for the restaurant or store. It can also be transformed to be a bedroom or a guestroom in order to make a suitable accommodation to the tourists or relative during the holiday. Besides, entertainment room can also be a good choice for the residents.

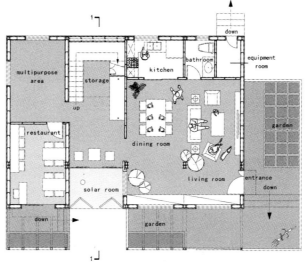

the second floor 1:50

the ground floor 1:50

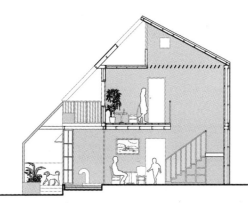

section 1 1:50

south facade 1:50

4278

Gardens in House 03
集院宅

Courtyard is set closely with living zone, like living room, bedroom, etc. The roof grills integrate with the garden, extending the entire yard from bottom to top. In addition to the ornamental values and comfortable living experience, the courtyard plays a significant role in passive energy-saving. Vines on the grills can provide shade. Modular planting can reduce the thermal radiation from the ground. Linear Stonecrop can insulate heat efficiently.

At day, green plants block the direct sunlight, which contains too much thermal radiation. Heated air is cooled by plants before entering into the rooms. At night, the faster heat dissipation of roof provides house owners with a cooling refreshing terrace.

At day, sunspaces make use of solar radiation to increase the interior temperature. Made up of PCM, the interior floor directly receive sunlight and store heat. At night, the PCM floor transits heat to the interior space, in order to reduce energy consumption of heating.

Courtyard Position — 1F Courtyard for farming — 2F Courtyard for landscape — 3F Courtyard for activity

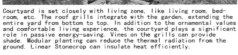

Function Position — 1F Hall & business — 2F Bedroom & tea room — 3F Roof terrace

Functional Envelope — Sunspace — PV panels — Courtyard

The staircase separate user-defined zone and living zone effectively so that it allows the dweller both allocate space freely and reduce the interference with the living zone. For example, if the user-defined zone is set to be a hotel or a restaurant, tourist activities will have little interference with the living zone.

Gardens in House 04
集院宅

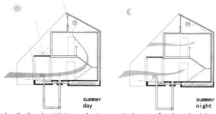

At day, the thermal ventilation accelerates convective heat transfer, the pool and the underground duct system send cool air into the room, and movable mats on the pool and attic shutters shade the sun effectively. At night, through-draught conduct convective heat transfer because of the casing window design. In addition, because the temperature decrease quickly in the roof, the cool air can enter underground gravel deposit through the northern rowlock wall, bringing out the residual heat storaged in the daytime.

At day, the sunspace and thermal-storaged tower of attic increase the room temperature by the sun radiation. The underground gravel deposit store the heat by the sun radiation and the hot air of the tower. At night, the side windows and shutters of the attic are closed so that the entire attic becomes a thermal insulation layer. Meanwhile, the underground gravel deposit releases the heat which was stored in the daytime.

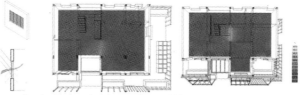

When using window shutters in the hot summer, the amount of light can be controlled and efficient ventilation can also be ensured. Interior daylight factors of all the areas will be greater than 2%, while the light will be mild and with small thermal radiation.

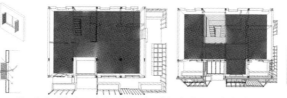

When not using window shutters in the cold winter, lots of thermal radiation transits into the rooms. Interior daylight factors of most areas will be greater than 2%, while the light will be adequate and with large thermal radiation.

Big width, small depth and appropriate windowing organize wind pressure ventilation effectively.

Annual incident solar radiation

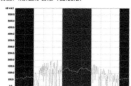

The passive energy-saving strategies need flexibility due to the climate features in Hot Summer and Cold Winter Zone.

Optimum orientation

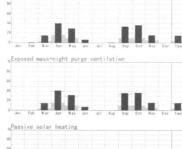

15 degrees east of south is the optimal orientation. Windows shading devices are necessary because of lots of yearly sun radiation in the east.

Passive energy-saving strategies

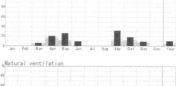

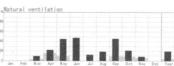

The priority order of passive energy-saving strategies:
Natural ventilation > Thermal mass effects > Exposed mass+night purge ventilation > Direct evaporative cooling > Passive solar heating > Indirect evaporative cooling

We decide the passive energy saving measures which is based on the building integration from the priority order of passive energy-saving strategies.

Gardens in House 05
集院宅

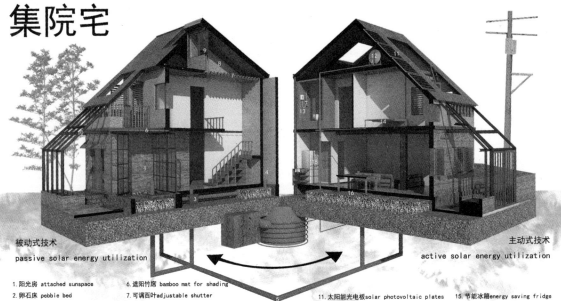

被动式技术 passive solar energy utilization

主动式技术 active solar energy utilization

1. 阳光房 attached sunspace
2. 卵石床 pebble bed
3. 活性炭 activated carbon
4. 空斗墙 rowlock wall
5. 地道通风管 underground vent pipe
6. 遮阳竹席 bamboo mat for shading
7. 可调百叶 adjustable shutter
8. 蓄热塔 solar chimney
9. 通风口 ventilation opening
10. 可拆卸式阳光棚 demountable solar pavilion
11. 太阳能光电板 solar photovoltaic plates
12. 蓄水箱 water storage tank
13. 节能热水器 energy saving water heater
14. 低温辐射床 thermal radiation bed
15. 节能冰箱 energy saving fridge
16. 热辐射地板 thermal radiation floor
17. 沼气池 biogas digester
18. 储气箱 biogas storage tank

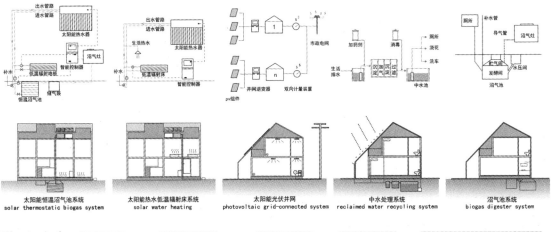

太阳能恒温沼气池系统 solar thermostatic biogas system | 太阳能热水低温辐射床系统 solar water heating | 太阳能光伏并网 photovoltaic grid-connected system | 中水处理系统 reclaimed water recycling system | 沼气池系统 biogas digester system

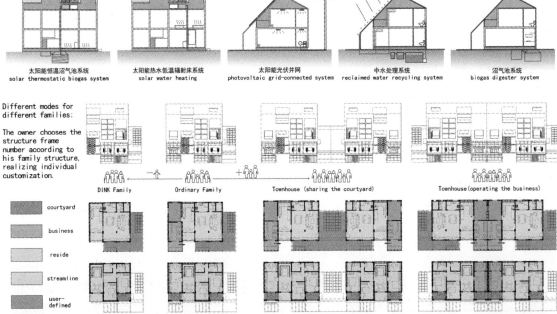

Different modes for different families:

The owner chooses the structure frame number according to his family structure, realizing individual customization.

DINK Family — Ordinary Family + Townhouse (sharing the courtyard) — Townhouse (operating the business)

- courtyard
- business
- reside
- streamline
- user-defined

Gardens in House 06
集院宅

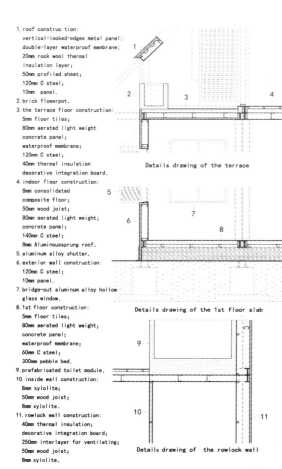
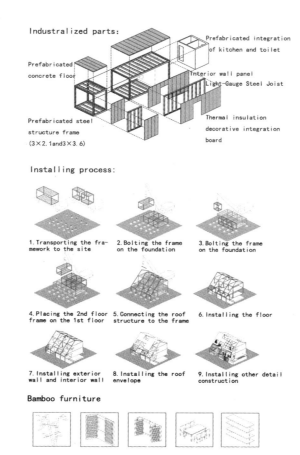
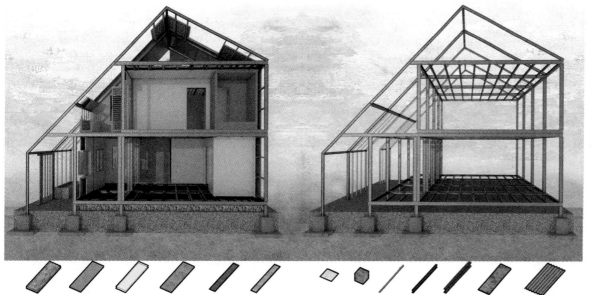

综合奖·优秀奖
General Prize Awarded ·
Honorable Mention Prize

注 册 号：4324
项目名称：耕·替（黄石）
　　　　　Transforming (Huangshi)
作　　者：易飞宇、丁中杉、黄 璐
参赛单位：东南大学

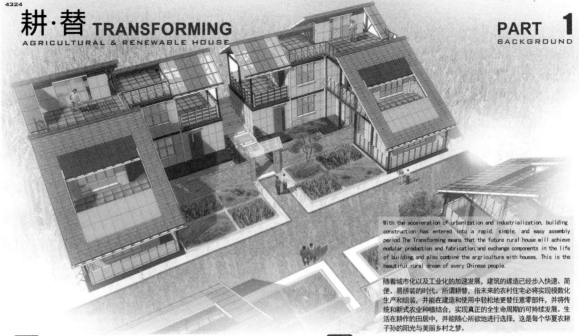

耕·替 TRANSFORMING
AGRICULTURAL & RENEWABLE HOUSE

PART **1**
BACKGROUND

With the acceleration of urbanization and industrialization, building construction has entered into a rapid, simple, and easy assembly period. The Transforming means that the future rural house will achieve modular production and fabrication, and exchange components in the life of building, and also combine the agriculture with houses. This is the beautiful rural dream of every Chinese people.

随着城市化以及工业化的加速发展，建筑的建造已经步入快速、简便、易拼装的时代。所谓耕替，指未来的农村住宅必将实现模数化生产和组装，并能在建造和使用中轻松地更替任意零部件，并将传统和新式农业种植结合，实现真正的全生命周期的可持续发展。生活在耕作的田居中，并能随心所欲地进行选择，这是每个华夏农耕子孙的阳光与美丽乡村之梦。

1.1 LOCATION AND CLIMATE ANALYSIS

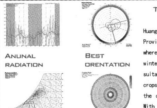

THE CLIMATE OF HUANGSHI

Huangshi is located in the Hubei Province, an abundant place, where summer is pretty hot and winter cold. The climate here is suitable for the growth of crops, but such weather reduces the comfort of living indoors. With the strong solar radiation and wind, here is a good place to develop solar architecture.

1.2 LIGHT ENVIRONMENT ANALYSIS

1.3 INDUSTRIALIZED HOUSE

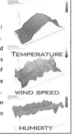

ORIGINAL HOUSE → INDUSTRIALIZED HOUSE

The traditional way of building houses uses materials directly. While the industrialized way is producing components in the factories.

A building can be broken into small elements according to the rooms or space.

THE ELEMENTS OF A HOUSE

PIECES

GROUPS

RECYCLE

These elements can be classified according to the location and size of the material, then preprocess elements in factory, and finally assembled.

1.4 THE DEVELOPMENT OF VILLEAGE

SCATTERED DISTRIBUTION → TREE SHAPE ORGANIZATION → HIGHLY ORGANIZATION

Rural area develops from the fragmented form to the tree form, then to a highly organized and planned shape. But at the same time the fields disappeared rapidly.

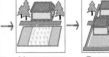

FORESTS　LOTS OF TREES　MANY TREES　FEW TREES

However, the expanding of cites and houses brings a huge destruction to forests. The ecological environment becomes bad. And the traditional industrilzation brings about more and more rural problems.

1.5 CURRENT RURAL PROBLEMS
LEFT-BEHIND YOUNG AND OLD

WEAKNESS LABOUR FORCE　ABANDONED FIELDS　CHANGE

耕·替 TRANSFORMING
AGRICULTURAL & RENEWABLE HOUSE

PART 2 MANUFACTURE

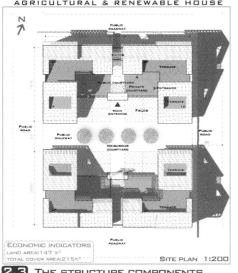

SITE PLAN 1:200

ECONOMIC INDICATORS
LAND AREA: 147 m²
TOTAL COVER AREA: 215 m²

2.1 THE COMBINATON OF COURTYARD

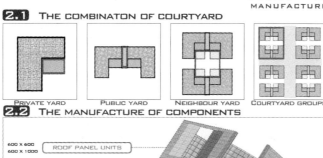

PRIVATE YARD | PUBLIC YARD | NEIGHBOUR YARD | COURTYARD GROUPS

2.2 THE MANUFACTURE OF COMPONENTS

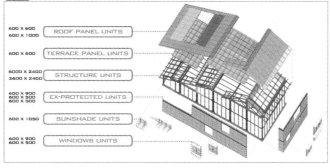

- 600 × 600 / 600 × 1000 — ROOF PANEL UNITS
- 600 × 600 — TERRACE PANEL UNITS
- 6000 × 2400 / 3600 × 2400 — STRUCTURE UNITS
- 600 × 900 / 600 × 500 / 600 × 500 — EX-PROTECTED UNITS
- 600 × 1050 — SUNSHADE UNITS
- 600 × 900 / 600 × 500 — WINDOWS UNITS

2.3 THE STRUCTURE COMPONENTS

STRUCTURE JOINT

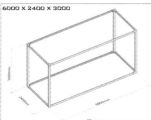
STRUCTURE UNIT 1 — 6000 × 2400 × 3000

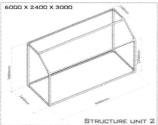
STRUCTURE UNIT 2 — 6000 × 2400 × 3000

STRUCTURE UNIT 3 — 3600 × 2400 × 3000

2.4 THE EXTERIOR-PROTECTED COMPONENTS

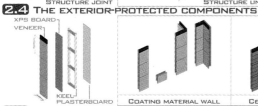
XPS BOARD / VENEER / KEEL / PLASTERBOARD

COATING MATERIAL WALL | CERAMIC PLATE WALL | PLASTIC STEEL WALL | WOOD PLATE WALL

2.5 THE ROOF PANEL COMPONENTS

ALTERABLE PLATES / LAY PLATE / FIXED FRAME / ROOF TRUSS

 TILE ROOF PLATE
 SOLAR HEATING PLATE
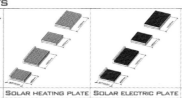 SOLAR ELECTRIC PLATE
 SKYLIGHT PLATE
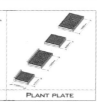 PLANT PLATE

2.6 THE DOORS AND WINDOWS COMPONENTS

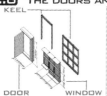
KEEL / DOOR / WINDOW

 DOOR-SOLIDWOOD
 DOOR-GLASS
 WINDOWSILL
 WINDOW-AWNING
 WINDOW-MIDDLE
 BAY WINDOWSILL

2.7 THE SUNSHADE COMPONENTS

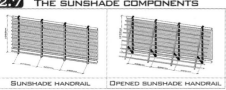
SUNSHADE HANDRAIL | OPENED SUNSHADE HANDRAIL

2.8 THE TERRACE PANEL COMPONENTS

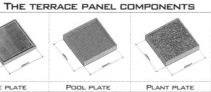
TILE PLATE | POOL PLATE | PLANT PLATE | MINI-FARM PLATE

耕·替 TRANSFORMING
AGRICULTURAL & RENEWABLE HOUSE

3.1 FABRICATION PROCEDURE

PART 3 — FABRICATION

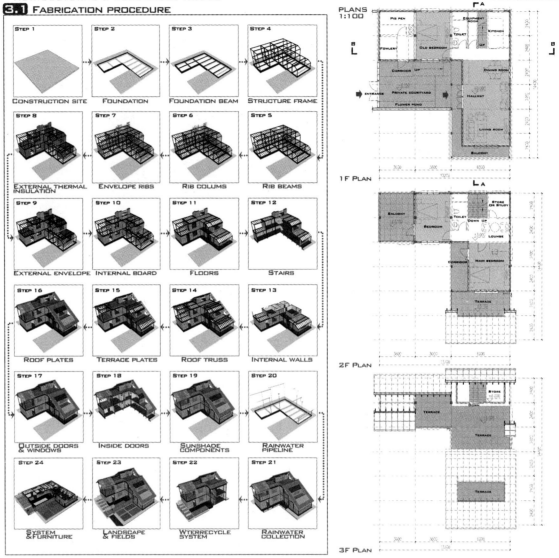

耕·替 TRANSFORMING
AGRICULTURAL & RENEWABLE HOUSE

PART 4 — USING & RECYCLING

4.1 THE RECYCLING CONDITION OF EACH SEASON & ALL LIFE OF BUILDING

Spring — Summer — Autumn — Winter

The components can be changed in every season and in each kind of situation. The walls and roof panels can be choosed by the owners and in order to their diffrent appettite, and some devices are designed to make the house more comfortable and humanity.

4.2 ALTERABLE HANDRAIL DEVICE

Spring & Winter Plant flowers

Summer & Autumn Sunshade

4.3 ALTERABLE VERTICAL CULTIVATE

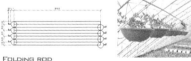

Folding rod

Folded condition — Unfolded condition

Cultivating trellis — Grape trellis

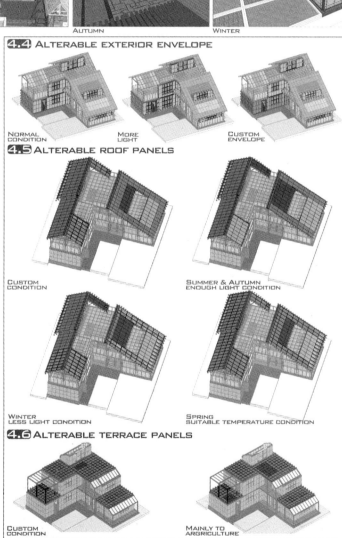

4.4 ALTERABLE EXTERIOR ENVELOPE

Normal condition — More light — Custom envelope

4.5 ALTERABLE ROOF PANELS

Custom condition — Summer & Autumn enough light condition

Winter less light condition — Spring suitable temperature condition

4.6 ALTERABLE TERRACE PANELS

Custom condition — Mainly to argriculture

South elevation 1:100 — West elevation 1:100

耕·替 TRANSFORMING
AGRICULTURAL & RENEWABLE HOUSE

PART 5
TRANSFORMATION

5.1 TRANSFORMATION OF SOLAR POWER

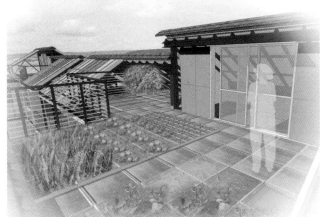
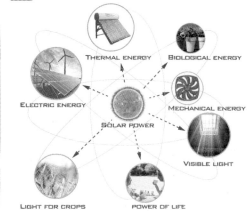

5.2 THE TRANSFORMATION FOR CULTIVATING

5.3 THE TRANSFORMATION OF AIR POWER

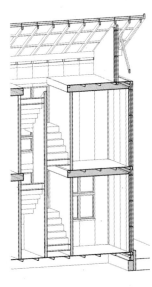
Ventilated hollow wall and roof

Winter day

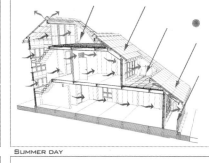
Summer day

Winter night

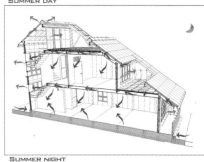
Summer night

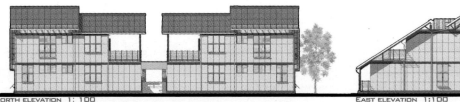

North elevation 1:100   East elevation 1:100

耕·替 TRANSFORMING
AGRICULTURAL & RENEWABLE HOUSE

PART **6**
TECHNOLOGY

6.1 THE USAGE OF WASTE MATERIALS

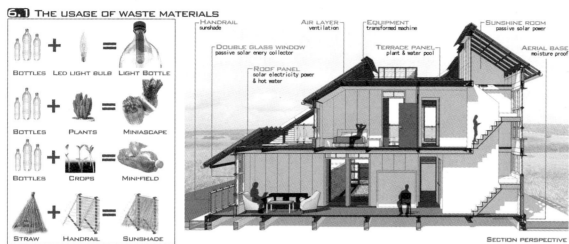

SECTION PERSPECTIVE

6.2 THE TREATMENT OF WASTE WATER & INTERMEDIATE WATER

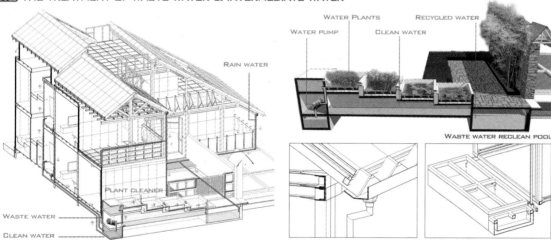

WATER RECYCLE SYSTEM · WASTE WATER RECLEAN POOL · RAIN WATER COLLECTION UNIT

6.3 THE TREATMENT OF RAIN WATER

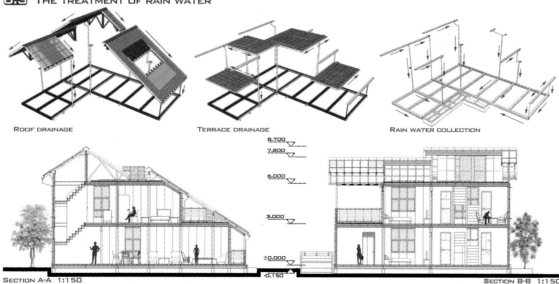

ROOF DRAINAGE · TERRACE DRAINAGE · RAIN WATER COLLECTION

SECTION A-A 1:150 · SECTION B-B 1:150

综合奖·优秀奖
General Prize Awarded · Honorable Mention Prize

注 册 号：4511
项目名称："风""光"无限好——美丽家乡（黄石）
Sunshine · Wind · Endless (Huangshi)
作　　者：刘　旭、张　静、夏　天、高才生、陈国锋、李博宇
参赛单位：南京工业大学

Sunshine · Wind · Endless

"风" "光" 无限好
——美丽家乡

Location analysis

The area climate of summer hot and winter cold is characterized by hot summers and winter cold, larger humidity of air, when the outdoor temperature is below 5 degrees Celsius, and if there is no heating, low indoor is lower temperatures, poor comfort. Huangshi city of the Valley to the East, West, North on three sides by mountains, and on the south side near the water, leading summer the Southeast wind, prevailing wind direction is to the Northwest in the winter, this is climate of regional climate models.

Valley is surrounded to mountains on three sides, the water opened to the south, have good lighting conditions, straightly into the summer of the southeast wind, have good ventilation. Winter winds obscured by mountains, location is good. Villages exists traditional street patterns and 2nd floor houses the main bay with 2,3. However, traditional design pattern is not good, north and south are not transparent, poor ventilation, sunlight will not be a good use of energy. And traditional ways of building construction need a long time, and building material waste is more serious, it have a greater impact on secondary use of building materials.

设计说明：
黄石市处于夏热冬冷地区，因为当地两开间和三开间的民居较多，我们组团也采用了这两种模式。我们规划的整体布局迎合了夏季东南风，北部山体阻挡冬季西北风，形成舒适小环境。每个组团都有南北两个公共的休闲广场。就单体而言，我们充分利用了光资源和风资源。单体建筑呈三段布置，中间形成空腔体系，冬季保暖，夏季拔风，与底层架空相结合。建筑前院的种植园，夏季利用垂直的植物遮阳，冬季覆盖上薄膜就是大棚，可以为室内供热。建筑之中还使用了阳光房、光电板、光热板等技术。建筑的外围护结构采用了外保温构造，屋顶采用了双层屋面。与此同时，建筑使用了便捷装配的轻钢结构，轻钢可以回收利用，绿色环保。

The city of Huangshi is very hot in summer and very cold in winter. Our group took the two modes which are the two standard rooms and the three standard rooms. The planning of this base caters to the southeast wind in summer and in winter the northern mountain obstructs the wind which could form a comfortable environment for the resident here. Each group of the houses has two public leisure square in the north and south. We make full use of the resources like sunshine, wind. The single building has 3 sections to arrange to form a cavity system in the middle of the house which supplys heat in winter to the surrounding space. In summer the cavity would combine with the bottom of the building. In front of the house there is a piece of plantation, in summer the vertical plant can form a shadow. In winter, it is covered with film just like green house to supply heat. Architecture also used the technology like the sun room, photovoltaic panels and photothermal plate. The peripheral structure of the building is used the external thermal insulation structure and the roof adopts double roof. At the same time, this building uses light steel structure. Steel is a material what can be recycled and be beneficial to the environment.

Page1 town planning

choose which structrue

Planning aerial view

Mapping and prototyping

Concept of single

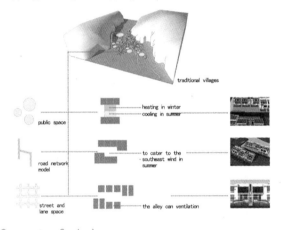

The space construct

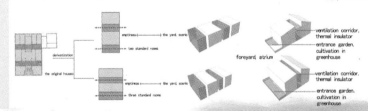

"风" "光" 无限好
——美丽家乡

The development of today's society is inseparable from the development of the structure of the building. With the development of technology, in the form of the building structure is also varied. However, in the form of traditional brick and concrete structure has, as shown on the left, construction waste pollution, dust pollution, high labor intensity, shockproof and poor performance of many shortcomings. The implementation of green housing industry is bound to explore the structure of the form. We can know from the table below, Web light-gauge steel structure system has excellent physical properties, while it can save manpower, recycle steels, reduce damage to the environment.

	Web light-gauge steel structure system	Steel structure	Reinforced concrete structure	Brick concrete structure
Safety	••••	••••	•••	••
Economy	•••	•••	••••	••••
Industrialization	•••••	•••	•••	•
Period	•••••	••••	•••	••
Workers	•••••	••••	••	••
Pollution	••••	••••	•	•
Limit	•••••	•••••	•••	••
Recycling	••••	••••	•	•

Note: More '•' respect superior and it is divided into 5-grades.

Analysis

The trend of distribution of population age Chinese diagram — From CNKI

Through the chart we can see that, with the passage of time, China's demographic dividend will be lost, labor population decline. Then China will enter the aging society, the urgent implementation of the process of industrialization.

Chinese urbanisation and income — From "people.cn"

Through the chart we can see that Chinese city urbanization continues and income level rises. So, Chinese labor's costs will increase in the future and it must be the trend of the future that machines replace people's hard work.

The chart of Chinese steel production and price — From "people.cn"

Through the chart we can see that output of crude iron rise and iron ore prices decline. From news, we know that steel is indisputable fact overcapacity. If we can spread this structure, it is helpful to consume surplus steels.

Sunshine · Wind · Endless

how to choose?

Trditional
V
S
Nowadays

I   II   III   V   VI

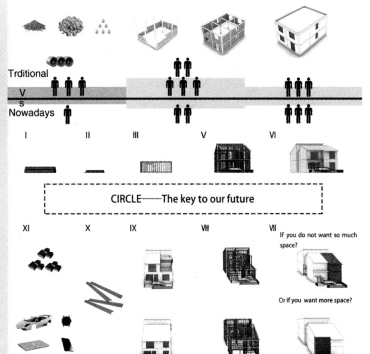

CIRCLE——The key to our future

XI   X   IX   VIII   VII

IF you do not want so much space?

Or if you want more space?

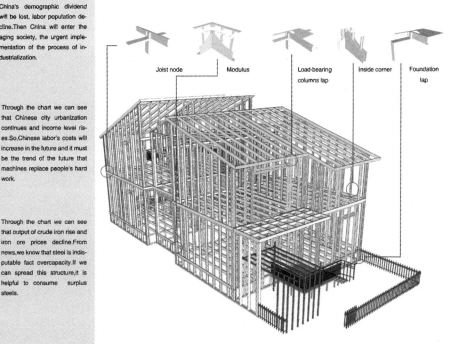

Joist node | Modulus | Load-bearing columns lap | Inside corner | Foundation lap

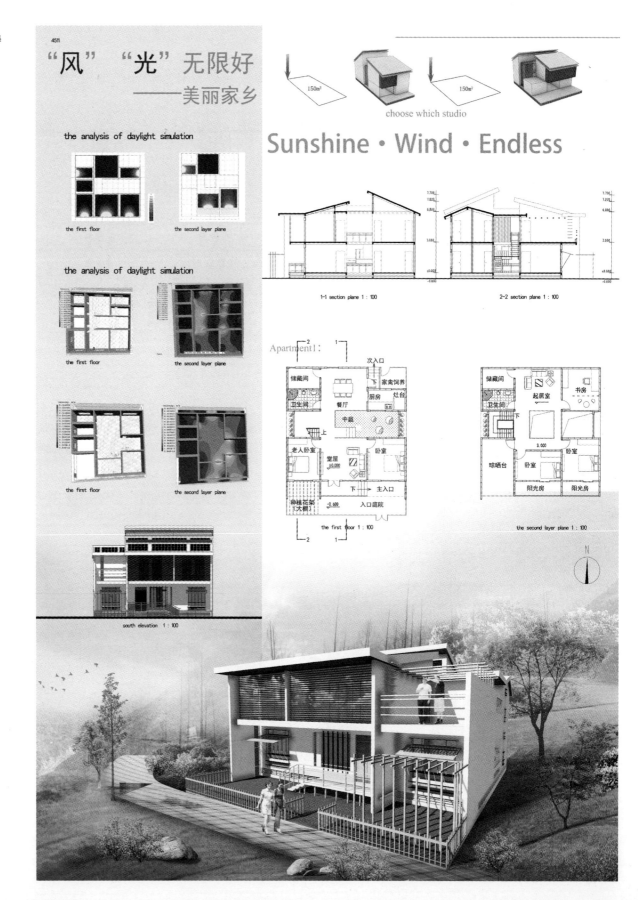

"风" "光" 无限好
——美丽家乡

Green technology analysis

This is a building block of the building block diagram, including both the function of the construction of the cavity body. Also according to the number of family members of the functional replacement concept, as well as the mobile space for architectural design.

Solar temperature room

Building replacement space

Indoor wind

Wind construction

Building overhead cavity

choose which studio

Sunshine · Wind · Endless

01. The analysis of builting on stilts

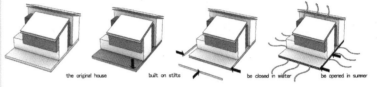

We will build the house overhead the ground about 0.6 meters, in winter it will be closed to reduce the loss of heat. In summer it will be opened. Then the wind through below it, and will remove the moisture.

02. The analysis of the movement of people

Building and building encloses the front and back of the square. In summer, people can play the activity in the shadow of the North space. In winter, people can enjoy the sunshine on the South Square.

03. The analysis of atrium

Architectural design can achieve the effect of saving energy in winter. South of the sun room and central atrium become a greenhouse, heating radiation into the surrounding space. In summer opening them and adding to the shading component, could contribute to the ventilation and take heat away.

04. The analysis of ventilation

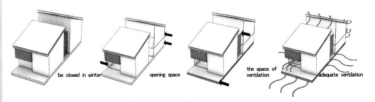

Wind is green energy. In summer, adequate ventilation, moisture could be removed and temperature can be dropped down.

"风" "光" 无限好
——美丽家乡

In the design of the solar house, we make fully the use of solar and wind energy, and strengthening the construction of residential thermal insulation and heat insulation itself. In the use of solar energy, we mainly use of passive solar energy utilization, such as solar house, heat insulation window and vegetable greenhouses. Active solar energy utilization includes photovoltaic system and solar thermal systems. In terms of shading is mainly building eaves shade and shutter shading. To solve the wet of Huangshi wet in summer, we use a building lower overhead approach, enhancing flooring-layer ventilation and dehumidification. Also, we have designed two profiles of summer and winter. Effect of summer through the Atrium stronger updraft, dehumidifying and heat insulation. Atrium has become a natural greenhouse in winter, delivering to heat on both sides of the room. Construction has two top roofs, one high and one low, low front and rear high, it is conducive to the summer windward and winter wind.

exploded axonometric

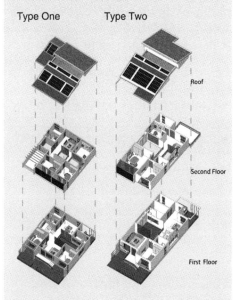

Type One · Type Two · Roof · Second Floor · First Floor

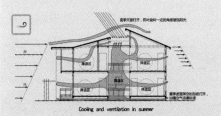
Cooling and ventilation in summer

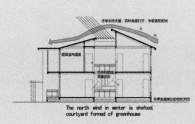
The north wind in winter is sheltered, courtyard formed of greenhouse

Page6 energy saving method

Sunshine · Wind · Endless
how to use them?

06. Solar house — summer / winter

07. Greenhouse — In the summer, the plant of pergola can block the sunshine. In winter, the greenhouse can provide heat for the interior.

08. Photovoltaic system

09. Solar panels

10. Two-double roof It can heat insulation.

11. Roof planting — It can reduce the urban heat island effect, increase air humidity and clean the air.

12. Energy-saving window — In winter day: Large windows closed, aboveand below the windows opened. In winter night: all of them shut down. In summer: all of them opened.

13. Louvres shading

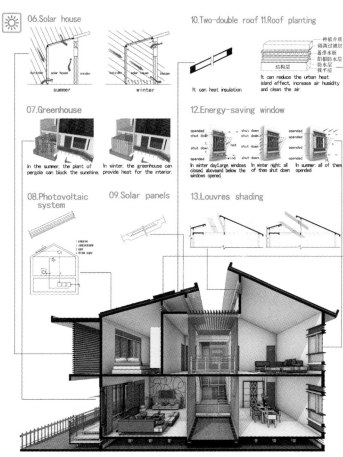

14. Septic tank — straw, faces, sewage entrance → the reaction → the residue is used as a fertilize; power homes exit

15. The treatment of rain and sewage — rain pipe

16. Tridimensional virescence — In summer the plants are flourishing to block sunshine. In winter the plants are withered and the wall is in the sunshine.

17. External wall thermal insulation structure — Gypsum board and steel are the main material composition of the envelope, and Huangshi is rich in the two kinds of building materials. Heat preservation material can prevent the indoor heat outside. The air infiltration layer can prevent outdoor thermal radiation in the summer.

18. "U" glass — "U" glass can form the air layer. In the summer, it can block the heat. In the winter, it can prevent bad weather and improve the indoor temperature.

有效作品参赛团队名单
Name List of all Participants Submitting Effective Works

注册号	作者	单位名称	指导人	单位名称
3299	张伟国、周腾飞、吴文超、叶佳贝、郑银炜	浙江大学宁波理工学院	周璟璟	浙江大学宁波理工学院
3301	刘红娟、许溪、吕怡卉、虞鸿飞、白雪峰、黄琳茹	广西大学	何江、倪轶兰	广西大学
3318	孟媛、芦浩、杨倩、解鹏昭、琼达	石家庄铁道大学、西藏大学	高力强、索朗白姆	石家庄铁道大学、西藏大学
3327	郑耀华、徐新衫、丁元、孔筱英、贾旭东	兰州理工大学、金陵科技学院、西安建筑科技大学华清学院	—	—
3330	符光伟、王晓娜	湖南城市学院	曾志伟、龙开宇、李佳伶、龚皓峰、刘沅	湖南城市学院
3343	杨帆、徐新杉、王洁、吴昇奕	金陵科技学院、南京大学	吴琅、戴军	金陵科技学院
3360	陈龙、周胜华、丁柳月、朱王倩、丛云天、冯硕、易强	西安科技大学	黄金诚	西安科技大学
3365	诸葛文斌、郭梦露	西安科技大学	孙倩倩	西安科技大学
3367	商选平	青海文旅投资有限公司	高庆龙、吴攀、陈瑞、商城毓	中国建筑西南设计研究院有限公司、青海文旅投资有限公司
3381	钟朗、黄翮、付一玲、蔡明倩	中央美术学院	崔鹏飞	中央美术学院
3397	董博文、曾晓丹、邢俊超	山东建筑大学、东南大学	房涛	山东建筑大学
3412	刘世良	河南省许昌市水利规划设计院	—	—
3413	Giacomo Liviabella、Anastasiu Bianca Iuliana、徐海洋	上海丝柏建筑设计有限公司	—	—
3422	张理奥、王嘉禾、郭畅、沈威	山东建筑大学	王江、管振忠	山东建筑大学
3432	肖菁羽、侯哲文、刘彬、钱佳庆	西安交通大学	陈洋、李志刚	西安交通大学
3437	李和勇、刘荣伶、费怡巍、王竞竞	河北工业大学、天津大学	舒平	河北工业大学
3438	姜黎、徐子、方亦媛	中央美术学院	苏勇	中央美术学院
3441	毛甜甜、李锦莉、金鹤年	中央美术学院	刘文豹	中央美术学院

续表

注册号	作者	单位名称	指导人	单位名称
3451	葛婧妍、郭皓月	中央美术学院	崔鹏飞	中央美术学院
3453	蒯新珏、刘琪睿	中央美术学院	崔鹏飞	中央美术学院
3456	艾洪祥、王洋、张天宇、郭冰月、赵珍仪、范斌、陈强、吕高标、宋薇、魏宏毫	山东建筑大学、重庆大学	薛一冰、杨倩苗	山东建筑大学
3460	姜旭	中央美术学院	何葳	中央美术学院
3461	周俊彤、王楚霄、王颖	中央美术学院	何葳	中央美术学院
3466	王浩坤、郭宇飞、刘明希	中央美术学院	何葳	中央美术学院
3468	吕佳依、刘名沛、苗九颖	中央美术学院	虞大鹏	中央美术学院
3470	刘子莘、赵今今、刘烨琳	中央美术学院	苏勇	中央美术学院
3472	胡云飞、梁欣、张兰	中央美术学院	—	中央美术学院
3474	李亚锦、王宗杰、马鑫	中央美术学院	虞大鹏	中央美术学院
3476	秦缅、隋昕、李宜轩	中央美术学院	虞大鹏	中央美术学院
3478	江衍璋、黄天植、王子健	中央美术学院	王环宇	中央美术学院
3493	徐慧敏、郑繁硕、鞠婧、徐睿鹏	山东建筑大学、Hochschule Anhalt (Bauhaus-University)	管振中、蔡洪斌	山东建筑大学
3509	韩念森、张泽茜、王惠、巴婧	天津大学、山东建筑大学	张玉坤	天津大学
3515	石文博、阿桂莲	昆明理工大学	吴志宏	昆明理工大学
3519	赵安国	山东建筑大学	—	—
3533	李航、卢杉	河北工业大学	舒平	河北工业大学
3534	刘成威、罗杰	西南交通大学	李百毅	西南交通大学
3538	陈超、闫佳佳、田甜	河北工业大学	舒平	河北工业大学
3551	赵普尧、陈永吉、焦宝峰、任艺梅、李妍妍、令宜凡	西安建筑科技大学	王军	西安建筑科技大学

续表

注册号	作者	单位名称	指导人	单位名称
3553	牛微、王楠、邵月婷、崔雅婧、孙楠、张玉洁、韩昆衡、赵若婉	山东建筑大学	薛一冰、杨倩苗、崔艳秋、刘长安、赵继龙、苗纪奎	山东建筑大学
3559	葛晓冰、辛尚、陈时洋、杨韵、衷逸群	重庆大学	周铁军、宗德新、张海滨	重庆大学
3571	李惠桦、晏凌峰、杜杨、周清华、吴少博、薛超、刘建松	北京交通大学	高巍、姜忆南、赵玫、杜晓辉	北京交通大学
3580	黄世华、邹汝波、杜文艺	昆明理工大学、重庆大学、华南理工大学	周伟	昆明有色冶金设计研究院
3585	李思超、殷梦芸	西安科技大学、西安建筑科技大学华清学院	—	—
3588	周晨、赵奕琳、刘诗柔、唐煜、俞彤霖	北京交通大学、汉能集团有限公司	曾忠忠、陈泳全、夏海山	北京交通大学
3597	商开洋、宋硕、张亚举	大连理工大学	胡英	大连理工大学
3598	宋硕、商开洋、张亚举	大连理工大学	胡英	大连理工大学
3603	李可欣、彭剑文、牛剑琨、万伟民、陈哲、李妍、朱维、蒋晓欣	湖北工业大学	张辉	湖北工业大学
3616	邵仕豪、杨松潮、陈佳丽	浙江理工大学	—	—
3617	李军迪、崔勋、朱集强	南阳理工学院	赵敬辛、谭征	南阳理工学院
3620	Yang Xu、Dipl.-Ing. Henning Drommer	IDS GmbH	—	—
3626	刘冲、姜羽平、胡春霞、梁一航、倪翰聪、杨晶晶、毕文蓓、张习龙	西安建筑科技大学	李钰、王军、闫增峰	西安建筑科技大学
3631	杜相、王强、胡达、吴昊、季京京、李欣	南京工业大学、合肥工业大学	胡振宇、丁炜	南京工业大学
3637	史纲纲、宛富强、谢海建	南阳理工学院	刘素芳、郑方园	南阳理工学院
3638	王轶楠、王冠宇、叶兆丹、闵韵然	重庆大学	周铁军、宗德新、张海滨	重庆大学
3650	孙德胜、赵文龙、柳恒松、李乐	南阳理工学院	刘素芳、郑方园	南阳理工学院
3658	林海、史振雷	复旦大学	陈月浩	复旦大学
3660	许泽寰、王雪菲、薛小刚、林蓓蓓	西安建筑科技大学、上海交通大学	雷振东	西安建筑科技大学

续表

注册号	作者	单位名称	指导人	单位名称
3664	曹世达	重庆师范大学	—	—
3676	周清华、吴少博、薛超、李楒桦、闫凌峰、杜杨、刘建松	北京交通大学	高巍、姜忆南、韩林飞、赵枚、杜晓辉	北京交通大学、北京理工大学
3677	高力强、加晶晶、仇朝兵、兰亮、庄冠存、刘其、李健、Jian Zuo	石家庄铁道大学、天津大学、University of South Australia	朱丽、孙勇	天津大学
3682	王竹林、刘莎、聂鑫、陈艳	沈阳工业大学	王海丹	沈阳工业大学
3687	邓伟艳、徐苇葭、牛梦乐、冯茜、荣琪、赵家旭、刘季渝	四川大学	张炜	四川大学
3689	康琪、吴翼飞、袁也、朱卓晖、高航、汪奥	湖北工业大学、深圳市同济人建筑设计有限公司	黄艳雁、张辉	湖北工业大学
3699	吴琼、赵延刚、叶雨辰、高力强、霍玉佼、刘其	天津大学	朱丽、孙勇	天津大学
3704	高力强、孟凡平、魏智强、张晶石、季思雨、尹欣、高洁、朱相栋	石家庄铁道大学、河北工业大学、清华大学	武勇、朱赛鸿	石家庄铁道大学、河北工业大学
3705	许锴	广州美术学院	梅策迎	广州美术学院
3712	钱世奇、沈宇驰、王建龙、刘洁莹	东南大学	钱强、徐小东、杨维菊、夏兵	东南大学
3715	张铮、赵一平、许蔓灵、莫旎卡、骆武辉、黄绮琪、郑楚烽、陈宗煌、关竣仁	华南理工大学建筑学院青年设计工作室	殷实	华南理工大学
3717	刘其、杨立博、高力强、赵延刚、吴琼、叶雨辰、李建	天津大学	朱丽、孙勇	天津大学
3720	甘月、王涵、左方圆、裴桂青、闫子川、欧阳松、梁晨、张晓寒	湖北工业大学	张辉	湖北工业大学
3723	尹欣、加晶晶、仇朝兵、陆文蕙、方东亚、侯丹蕾	石家庄铁道大学、河北建筑工程学院	高力强、郭晓君	石家庄铁道大学、河北建筑工程学院
3724	苏珊、朱赛鸿	河北工业大学	—	—

续表

注册号	作者	单位名称	指导人	单位名称
3729	李非、刘晨芳、刘宇文、颜博达、钟成	北京交通大学	杜晓辉	北京交通大学
3732	王天亿、王磊	河北工业大学、华优建筑设计院	朱赛鸿	河北工业大学
3734	苏俊霆、段又榕、车晓路、卢小红	吉林建筑大学	李佳艺	吉林建筑大学
3737	傅嘉言、孙姣姣	浙江大学	王竹、罗卿平	浙江大学
3740	霍玉佼、邵泽彪、吴琼、赵延刚、刘其、李健	天津大学	朱丽、孙勇、赵旭	天津大学、中澳城市环境与可持续发展研究中心
3747	兰亮、庄冠存、孟凡平、魏智强、Jian Zuo、韩涛	石家庄铁道大学、University of South Australia、天津大学	高力强、武勇	石家庄铁道大学
3755	温雯、李煜茜、刘亚彬、张亚琼	北京交通大学	石克辉、杜晓辉	北京交通大学
3764	陈迪、宦烨晨、李辉、王嘉萌、原宇、杨宏博	西安建筑科技大学	王军、靳亦冰	西安建筑科技大学
3768	安娜、魏宏毫、张晨悦、提姗姗、邢龙飞	山东建筑大学	薛一冰、杨倩苗	山东建筑大学
3769	綦岳、张英、商宇辰	山东建筑大学	薛一冰	山东建筑大学
3774	汪昇、李云飞	南京工业大学	蒋晓风	南京工业大学
3778	周世杰、宋振威、庞树林、郭良坤、胡冬冬、刘艳丽、沙媛莉、龙垣屹、毛杰、周云龙	滨州学院、东南大学、CCDI悉地国际建筑设计顾问有限公司、江苏福万家新能源科技有限公司	杜婷、李慧、陈璐	滨州学院
3781	颜会间、汤蓓、刘雪、罗侬侬、齐成、鲍明、周青	武汉大学	宋靖华、王炎松	武汉大学
3782	谭子龙、徐少敏、杨玉锦	南京大学	—	—
3787	王文祥、路程	西安科技大学	—	—
3806	葛贵武、刘宁波、李娴、朱宸、梅岩	解放军后勤工程学院、重庆市轻工业学校	—	—
3809	李童、鲁会凯、徐莉、宋书剑、王嘉凯	山东建筑大学	李晓东	山东建筑大学

续表

注册号	作者	单位名称	指导人	单位名称
3813	蔡雨馨、陈晓月、程子珊、明阳、南漪、任嘉友	四川大学	张炜、陈红	四川大学
3816	张浩然、刘哲迪、刘思源、战凯、李清晴	四川大学	吕思强	四川大学
3826	石成、丁和	暨南大学、山东建筑大学	何文晶	山东建筑大学
3828	张良钰、李佳佳、肖华杰	东南大学	杨维菊	东南大学
3830	李菲菲、赵雅威、洪宇东、徐申	北京交通大学	石克辉、杜晓辉	北京交通大学
3832	徐金龙、姚成、刘茵、徐金虎、李瑞、何建涛、张鹏、李家翔	西安建筑科技大学、Heriot Watt University	杨柳、沈西平、Josh Ng	西安建筑科技大学、西安建筑科技大学建筑设计院、Heriot Watt University (UK)
3834	安南旭、郑国澎、冯量、张旭	中国中建设计集团有限公司	—	—
3837	林海锐、刘小康、胡浩森、何傲天、刘晶、周楚晗、朱嘉懿、许安江、周一冲	华南理工大学建筑学院青年设计工作室	钟冠球	华南理工大学
3838	魏易盟、李陶、杨明慧、王逸凡、张泽华、何世群、刘彤	青岛理工大学	郝赤彪、解旭东、程然、耿雪川	青岛理工大学
3847	骆肇阳	贵州大学	—	—
3865	高思晗、易柯辛	中国人民大学	—	—
3866	张靓秋、张保国	佚名建筑工作室(香港)、厦门瑞景景观设计公司	—	—
3869	王进、黄婷、朱文瀚、仇银涛、钟立涵、陈宇、杨晨	盐城工学院	—	—
3874	童威、张成、童飞	南阳理工学院、天津城建大学	刘素芳、郑方园	南阳理工学院
3881	侯喆文、靳湾湾、钱宗祎	南京工业大学	胡振宇	南京工业大学
3887	李青、郑泽闽、邓成文、邓成柳、包雪、周旋、李易斌、代运成	湖北工业大学	张辉	湖北工业大学

续表

注册号	作者	单位名称	指导人	单位名称
3891	杨三瑶、江捷椿、刘飒沙、刘家旺、纪圣霖、许梦林、凌倩瑜、张津	合肥工业大学	凌峰	合肥工业大学
3896	张翔、邵京金	北京交通大学、University of Sheffield	—	—
3898	杨一苇、谢鹤男、张睿智、李瑞琪、贺鑫、陈勇、王松、荣琪	四川大学	张炜、林野	四川大学
3902	赵泽祥、毕雨晨	滨州学院	—	—
3903	王鑫、梁西、黄冠道、曹恺悦	北京交通大学	王鑫（老师）、杜晓辉	北京交通大学
3907	倪芸、鲍纬哲	南京工业大学	胡振宇	南京工业大学
3912	王春彧、李昊、张林凝、张煦康、李昆、丁仕琪、马楠、李瑞	武汉大学	黄凌江	武汉大学
3921	侯信杰、马子茹、孟媛、芦浩、王晓静、王冰华	石家庄铁道大学、天津大学	高力强、朱丽	石家庄铁道大学、天津大学
3923	杨倩、解鹏昭、马子茹、侯信杰、丹增索朗	石家庄铁道大学、西藏大学	高力强、索朗白姆	石家庄铁道大学、西藏大学
3925	刘莉、徐潜	南京工业大学	蔡志昶	南京工业大学
3927	王玮颉、韩叙、王越、宝一力、贾琼、刘彦麟	同济大学	—	—
3932	孙恩格、姚竹西、钱昆、孙思雨	武汉大学	黄凌江	武汉大学
3935	黄滢、洪安东、郁晶	淮阴工学院	胡海洪、康景润	淮阴工学院
3937	陈强、袁军、徐凯、宋帅、宋薇	山东建筑大学	薛一冰、杨倩苗	山东建筑大学
3948	徐树刚、郑浩	深圳大学	吴向阳	深圳大学
3951	仲亮、陈子嘉、任宇	西南交通大学	李百毅、左辅强	西南交通大学
3956	辛莘、李牧笛、陶云帆	中国人民大学附属中学	施一宁	中国人民大学附属中学
3963	方东亚、侯丹蕾、丁磊、李佳琳、王岚、韩涛	石家庄铁道大学、天津大学	高力强	石家庄铁道大学

续表

注册号	作者	单位名称	指导人	单位名称
3967	鲁承文、高瑞仓、刘爽、邵琦、何冬梅、于添胤、洪群凯、莫福祥、贾峰、张津铭、郭璧圣、梁瀚	东北石油大学	马令勇、张天宇、任洪国、徐晓丽	—
3969	张净妮、刘真、佘潇、高寒钰、邱建维、梁伟杰、姜少华	四川大学、同济大学	李强、张炜	四川大学
3971	朱明、宋雨斐、王洋	山东建筑大学	薛一冰	山东建筑大学
3972	高深、张振、贾峰、张信雅、陈晓东、程子瑜、杨镇、高晶、张建凤	吉林建筑大学、东北石油大学	张萌、周洪涛、马令勇、赵文艳、任洪国、孟子南、王科奇、柳红明、李之吉	吉林建筑大学、东北石油大学
3983	陶成林、柳鹏博、田启晶、陈柏宇	青海建筑职业技术学院	王延存、王刚	青海建筑职业技术学院
3985	庄梓涛、肖泽恒、余光鑫、张贵彬、王良亮、李张君、刘智伟、李玥	华南理工大学建筑学院青年设计工作室	王静	华南理工大学建筑设计研究院、华南理工大学
3995	王雅、王德玲	中南建筑设计院股份有限公司	—	中南建筑设计院股份有限公司
3996	王森、尹家欢、刘一君、星妍、黄丽嬗	北京交通大学	杜晓辉、曾忠忠	北京交通大学
3997	王博伦、徐佳臻、李震寰、李维珊、葛小蓉、谷红磊	大连理工大学	刘鸣	大连理工大学
3998	梁栋、赵思源、李霭峰、薛骋、王隆勋、赵鸣、徐薇淼、徐洋	华中科技大学	刘晖、李保峰	华中科技大学
4002	葛倩、胡亚男、章登、孙娱、兰彤	华中科技大学	余庄	华中科技大学
4006	胡亦夏、高隽璐、鲜卓麟、王旭弟	重庆大学	周铁军、宗德新、张海滨、覃琳	重庆大学
4012	陈高涛、季莉莉、肖璇、罗西子、冯昱	重庆大学	周铁军、宗德新、张海滨	重庆大学
4019	江璇、周筱扬、张蕊、杨元元	西南交通大学	—	—
4022	林嘉娜、李思颖、付叶银、刘彤凯、孟宪宏、蒋卓洵、覃涵、简祎	四川大学	李炜	四川大学

续表

注册号	作者	单位名称	指导人	单位名称
4025	朱敦煌、黄晨虹、肖迦煜、于皓琳、肖玥	武汉大学	李鹃	武汉大学
4029	李新宇、王晓青、王金达	山东建筑大学	任震	山东建筑大学
4033	沈梦乔、李佳威、李硕	南京工业大学、The University of Texas at Austin	薛春霖	南京工业大学
4045	曹乔乔、李超、陈健、刘茂刚	吉林建筑大学	张广平	吉林建筑大学
4046	胡振宇、邵继中、戴天序、马强	南京工业大学	—	—
4047	周振文、张逸丰、冯智渊、韩雪、靳昕	华北水利水电大学	高长征、董姝婧、张东、王桂秀、卢玫珺、宋海静	华北水利水电大学
4048	张振、陈晓东、贾峰、张信雅、高深、程子瑜、杨镇、张建凤	吉林建筑大学、东北石油大学	周洪涛、王科奇、马令勇、赵文艳、任洪国、孟子南、张萌、李之吉	吉林建筑大学、东北石油大学
4049	尹梦泽、刘海婧、王骥、段飞、迟天庆、蓝亦睿、鹿少博、张烨	山东建筑大学	仝辉、赵斌、李晓东	山东建筑大学
4054	刘宇、王玉婕、桑雨岑、高小燕	重庆大学	周铁军、宗德新、张海滨	重庆大学
4059	郭乐乐、蔡变蓉、胡振宇	南京工业大学	—	—
4062	黄习习、孙然、郑洁、王艳雪、杨传真、陆曦	山东建筑大学	任震、郑恒祥	山东建筑大学
4066	纵凯、谭跃、周伟、吴冬乐、王喆	盐城工学院	王进、黄婷	盐城工学院
4068	侍帅、叶凌峰、肖强、顾峰、赵宸	盐城工学院	王进、黄婷	盐城工学院
4069	葛睿、吴佳沁、王月花、黄康民、王海英	盐城工学院	王进、黄婷	盐城工学院
4070	姜广君、华露嵘、单铭、罗立杰、彭鹏、高语晗	盐城工学院	王进、黄婷	盐城工学院
4071	刘冠婕、仲文娟、崔岭、陈佳慧、陶璟汶	盐城工学院	王进、黄婷	盐城工学院
4079	吴漫意	东华大学	黄更	东华大学

续表

注册号	作者	单位名称	指导人	单位名称
4087	杨鸿玮、毕晓健	天津大学	刘丛红	天津大学
4088	郭畅	山东建筑大学	王江、管振忠	山东建筑大学
4091	马小娟、袁必富、陈峰	南京工业大学	胡振宇、张伟郁	南京工业大学
4092	鲍纬哲、倪芸	南京工业大学	胡振宇	南京工业大学
4108	蔡子君、陈耀芳、林洁、赖祥助	浙江农林大学	何礼平、王美燕	浙江农林大学
4110	曹菁菁、陈昊	南京工业大学	丁炜、蒋晓峰	南京工业大学
4112	Miguel Cruz、Stefan Jovanovic、Elita Medina	重庆大学	周铁军、宗德新、张海滨	重庆大学
4114	程雪杨	安阳工学院	张仲军	安阳工学院
4118	卢森茂、孙思远	南京工业大学	蒋晓风	南京工业大学
4125	丁中杉、马迅、黄璐、陈文强、易飞宇	东南大学	彭昌海、万邦伟	东南大学
4127	阎智源、李慧聪	南阳理工学院	赵敬辛、谭征	南阳理工学院
4132	于卓群、闫倩	吉林建筑大学	柳红明、周春艳	吉林建筑大学
4135	郭一萌、刘昱彤	南京工业大学	蒋晓风	南京工业大学
4141	吴昌亮、孙杰、徐亮、徐斌、李哲健、刘大用、马镇宇	东南大学、汉能全球光伏应用集团	杨维菊、陈文华	东南大学、汉能全球光伏应用集团
4142	刘大用、马镇宇、王晨杨、吴昌亮、孙杰、徐亮	东南大学、汉能全球光伏应用集团	杨维菊、万邦伟、陈文华	东南大学、汉能全球光伏应用集团
4150	王丹丹、冯晓骅	南京工业大学	丁炜、蒋晓峰	南京工业大学
4159	张宁、冯伟杰、魏娜、王云	大连理工大学	范悦、李国鹏	大连理工大学
4166	李立、曾宸、张岱宗、张灿、伹颖鑫、陈佳然	北京交通大学	陈泳全、曾忠忠、夏海山	北京交通大学
4167	曹菁菁、陈昊	南京工业大学	丁炜、蒋晓峰	南京工业大学

续表

注册号	作者	单位名称	指导人	单位名称
4170	雷鑫、钱宇辰	南京工业大学	丁炜	南京工业大学
4175	蔡利媛、高贞、陈佳灯、欧雷	南京工业大学	欧雷	南京工业大学
4180	王露云、王恺成、张蕾、周胤鹏	南京工业大学	丁炜	南京工业大学
4183	黄杰、吴鑫澜、韩凯、陈诗、张嫩江、宋祥	西北工业大学、西安建筑科技大学	刘煜、王军	西北工业大学、西安建筑科技大学
4186	王云朋、李加志、钟国栋、龙瞬杰、杨基伟	华中科技大学	余庄	华中科技大学
4187	张然、邢健、高宏强、邹志鹏	合肥工业大学	饶永	合肥工业大学
4192	孙睿、朱杰	南京工业大学	蔡志昶	南京工业大学
4199	罗星、易柏存、张波、魏维军	攀枝花学院	马瑞华	攀枝花学院
4200	孙一富、刘阳、陈菲	内蒙古科技大学	马明	内蒙古科技大学
4203	赵殷英	武汉理工大设计研究院有限公司	—	—
4220	郑斯宇、刘航、凌捷、王逸豪、徐建欣	盐城工学院	王进、黄婷	盐城工学院
4221	邓健聪	广州美术学院	—	—
4228	李江华	江苏科技大学	翟晓婷、冉祥辰、董悦尧、胡延仁	江苏科技大学
4229	侯坤奇、许岭、李艳霞	东南大学	杨维菊	东南大学
4231	王悦、李晓阳、金潇雪	大连理工大学	范悦、李国鹏	大连理工大学
4236	Ahmed Mohamed Imam	Cairo University	—	—
4239	曾晓丹、顾容竹、侯晓、李英璞	山东建筑大学	房涛	山东建筑大学
4241	应振国、曹世彪、王晓楠、诸葛涌涛	天津大学、无唯工作室	贾巍杨	天津大学
4246	车成业、李润东、王子晗、袁哲	东北大学	—	—
4256	贺靖、杨慧琳、李嘉慧、曾宸	北京交通大学	曾忠忠、陈泳全、夏海山	北京交通大学
4258	涂炜玮、李沁霏、房盼、朱虹旭	江苏科技大学	任鹏远、陈萍	江苏科技大学

续表

注册号	作者	单位名称	指导人	单位名称
4262	黄婷婷、黄文佳	南京工业大学	蒋晓风	南京工业大学
4263	季思雨、张经纬、吴迪、张晶石、高洁、高力强、郑铮	河北工业大学、天津大学	朱赛鸿、王朝红	河北工业大学
4265	陈慧贤、朱婷、邵继中	南京工业大学	—	—
4266	刘罡、刘紫伊、梁鑫、邵继中	南京工业大学	—	—
4276	李根、唐滔、卢莫瑞、刘丹、李立	北京交通大学	薛彦波	北京交通大学
4278	黄绮琪、郑楚烽、陈宗煌、关竣仁、张铮、赵一平、许蔓灵、莫旎卡、骆武辉	华南理工大学建筑学院青年设计工作室	林翰坤	华南理工大学
4289	谈为康、范艳、朱安平、王成	江苏科技大学	韦薇	江苏科技大学
4299	袁磊、刘怡斐	清华大学、天津美术学院	彭军、方晓风	天津美术学院、清华大学
4305	Karina Zanon	Mackenzie Presbyterian University	Carlos Andrés Hernandéz Arriagada	Mackenzie Presbyterian University
4307	刘欢、李扬淑、徐弋、高雪辉	天津大学	贾巍扬	天津大学
4312	侯亚松	南阳理工学院	赵敬辛、谭征	南阳理工学院
4324	易飞宇、丁中杉、黄璐	东南大学	彭昌海、万邦伟	东南大学
4330	徐宁、农崟荷、李湉、王文静、陈未	北京工业大学	陈喆	北京工业大学
4335	锡望、陈亚童、潘玥、杨婧一、高焱	北京建筑大学	李英、邹越	北京建筑大学
4344	郭欣雨、胡晓蔚	南京工业大学	欧雷	南京工业大学
4346	申强、丁亚萍、郭缓	黄淮学院	张晓峰	黄淮学院
4347	胡琪蔓、向奕妍、张璐、冯昱、杨巧霞	重庆大学	周铁军、宗德新、张海滨	重庆大学
4348	徐梓斐、侯义	河南大学	—	—
4349	尹博闻、吴雪杰、周师	西南科技大学、湖南文理学院	赵祥	西南科技大学

续表

注册号	作者	单位名称	指导人	单位名称
4355	王欣、卫泽华、樊梦蝶、荆济鹏、张奕聪	北京交通大学	陈泳全、曾忠忠、夏海山	北京交通大学
4372	陈莉、柴克非、李垚、薛凯	重庆大学	周铁军、宗德新、张海滨	重庆大学
4373	陈维发、鄂菲、巩何杉、陈思蓉、李祥瑞、荆文强	中国中铁二局集团有限公司、西安建筑科技大学	—	—
4382	胡进玮、曹安琪、李蕾、胡露	华中科技大学	—	—
4388	姜哲、杜雪、周驰、张皓然、张展翔	中国矿业大学（北京）	李晓丹	中国矿业大学（北京）
4390	李建、岑超、李东遥、高力强	天津大学	朱丽、孙勇	天津大学
4396	刘圣泽、胡晓婷、陈玉婷	哈尔滨工业大学	孙清军、吴健梅	哈尔滨工业大学
4397	石刘睿恬、张军军、郭书凯	东南大学	张宏、傅秀章、张弦	东南大学
4398	张军军、孔亦明、石刘睿恬、郭书恺	东南大学	张宏、傅秀章、张弦	东南大学
4399	郭书恺、石刘睿恬、张军军	东南大学	张宏、傅秀章、张弦	东南大学
4411	尹梦泽、刘海婧、王骥、段飞、迟天庆、蓝亦睿、鹿少博、张烨	山东建筑大学	仝辉、薛一冰、赵斌	山东建筑大学
4420	张鹏、刘轩、李芸、邹清妍	华北水利水电大学	卢玫珺、王桂秀、高长征、张东、董姝婧	华北水利水电大学
4423	李翊、赵金剑、高安妮、张朝虎、陈琨、吴银光、杨雷	中南建筑设计院股份有限公司	—	—
4439	丁磊、陈思源、张文、张晓斐	天津大学、青岛绿城建筑设计有限公司	郭娟丽	天津大学
4444	刘枫、陈敏、刘袁芳、蔡伟	华中科技大学	—	—
4445	董吉程、朱杰、王嘉威、薛晨	南京工业大学	欧雷	—
4461	吴琦炜、许永超、陈磊	浙江大学	朱伟	浙江大学
4466	孙姣姣、傅嘉言	浙江大学	贺勇	浙江大学
4469	曾世吉、李伟、彭明明、钟振宇、陈越、刘丽珍、任瑞雪	南昌大学	徐从淮	南昌大学

续表

注册号	作者	单位名称	指导人	单位名称
4474	翁之韵、徐宛清、熊申午、陶鹏飞	南昌大学	徐从淮、魏永健、郑文晖	南昌大学
4476	范恺、罗璇、钟楷	南昌大学	徐从淮、魏永健	南昌大学
4477	于雪、项琛春、胡婧坤	南昌大学	徐从淮、魏永健	南昌大学
4478	赵期文、徐蜀辰、张然、陈越	南昌大学、同济大学、广州宝贤华瀚建筑设计有限公司南昌分公司	徐从淮、魏永健、郑文晖	南昌大学
4483	陈景晔、李志勤、师自杰、张东、王桂秀、卢玫珺、高长征、董姝婧、宋海静	华北水利水电大学	—	—
4485	黄誉、李岩、肖志彪、陈聪睿	南昌大学	徐从淮	南昌大学
4486	王冠军、吴一姣、张宇涛、邵继中	南京工业大学	—	南京工业大学
4497	陈倩仪、王熹、王怡静	华南理工大学	郭明卓	广州市设计院
4500	黄辉、黄思源	南京工业大学	蔡志昶	南京工业大学
4511	刘旭、张静、夏天、高才生、陈国锋、李博宇	南京工业大学	钱才云、张伟郁、周扬	南京工业大学
4513	吴晓凡	山东建筑大学	金文妍	山东建筑大学
4516	Chiara Luchino	Sapyenza University of Rome	—	—
4517	林显、刘天然、杨骏卿、姚远、张莞晨、商城毓、金鑫	西安建筑科技大学	靳亦冰、商选平	西安建筑科技大学、陕西省建筑设计研究院
4518	Jorge Reynaud	University of Montreal	Roger-Bruno Richard	University of Montreal
4521	Gabrielle Charpentier、Pascale Ducas	University of Montreal	Roger-Bruno Richard	University of Montreal
4522	Fêten Magri-Foughali	University of Montreal	Roger-Bruno Richard	University of Montreal
4525	郭笑晨、朱婷婷	重庆大学	王雪松	重庆大学

2015台达杯国际太阳能建筑设计竞赛办法
Competition Brief for International Solar Building Design Competition 2015

竞赛宗旨：

在建设美丽乡村和推动建筑工业化进程中，努力实践太阳能利用等绿色、低碳、健康技术，研发经济型宜居农村住房。

竞赛主题：阳光与美丽乡村

竞赛题目：农牧民定居青海低能耗住房项目
农村住房产业化黄石住宅公园项目

主办单位：国际太阳能学会
中国可再生能源学会

承办单位：国家住宅与居住环境工程技术研究中心
中国可再生能源学会太阳能建筑专业委员会

支持单位：中华人民共和国住房和城乡建设部建筑节能与科技司
中国建筑设计研究院

冠名单位：台达环境与教育基金会

评委会专家：崔愷：国际建协竞赛委员会委员、中国建筑学会常务理事、中国工程院院士、中国建筑设计院有限公司名誉院长、总建筑师。
Anne Grete Hestnes 女士：前国际太阳能学会主席，挪威科技大学建筑系教授。
Deo Prasad：国际太阳能学会亚太区主席，澳大利亚新南威尔士大学建筑环境系教授。
M.Norbert Fisch：德国不伦瑞克理工大学教授（TU Braunschweig），建筑与太阳能技术学院院长。

GOAL OF COMPETITION:

In the process of building beautiful villages and promoting industrialization, it is necessary to apply renewable resources such as solar energy, and green and low carbon building technology in pursuit of affordable livable rural housing.

THEME OF COMPETITION:

SUNSHINE AND BEAUTIFUL VILLAGE

SUBJECT OF COMPETITION:

Subject I: Low Energy-Consumption Housing for Farmers in Qinghai Province
Subject II: Rural Housing Industrialization in the Residential Park, Huangshi City

ORGANIZER:

International Solar Energy Society (ISES)
Chinese Renewable Energy Society (CRES)

OPERATOR:

China National Engineering Research Center for Human Settlements (CNERCHS)
Special Committee of Solar Buildings, CRES

SUPPORTED:

Department of Building Energy Saving and Science and Technology, Ministry of Housing and Urban-Rural Development of the People's Republic of China
China Architecture Design & Research Group

Peter Luscuere：荷兰代尔夫特大学（TU Delft）建筑系教授。

Mitsuhiro Udagawa：国际太阳能学会日本区主席，日本工学院大学建筑系教授。

林宪德：台湾绿色建筑委员会主席，台湾成功大学建筑系教授。

仲继寿：中国可再生能源学会太阳能建筑专业委员会主任委员，国家住宅与居住环境工程技术研究中心主任。

喜文华：联合国工业发展组织国际太阳能技术促进转让中心主任，联合国可再生能源国际专家，国际协调员，甘肃自然能源研究所所长。

黄秋平：上海现代设计集团华东建筑设计研究院副总建筑师。

冯雅：中国建筑学会建筑热工与节能专业委员会副主任，中国建筑西南设计研究院副总工程师。

组委会成员：由主办单位、承办单位及冠名单位相关人员组成。办事机构设在中国可再生能源学会太阳能建筑专业委员会。

评比办法：

1. 由组委会审查参赛资格，并确定入围作品。
2. 由评委会评选出竞赛获奖作品。

评比标准：

1. 参赛作品须符合本竞赛"作品要求"的内容。
2. 作品应具有原创性和前瞻性，鼓励创新。
3. 作品应满足使用功能、绿色低碳、安全健康的要求，建筑技术与太阳能利用技术具有适配性。
4. 作品应充分体现太阳能利用技术对降低建筑使用能耗的作用，在经济、技术层面具有可实施性。
5. 作品评定采用百分制，分项分值见下表：

SPONSOR:

Delta Environmental & Educational Foundation

COORGANIZER:

Architecture Technique

EXPERTS OF JUDGING PANEL:

Mr. Cui Kai, Commissioner of International Union of Architects, Standing Director of Architectural Society of China, Academy of China Academy of Engineering, Honorary President and Chief Architect of China Architecture Design Group.

Ms. Anne Grete Hestnes, Former President of International Solar Energy Society and Professor of Department of Architecture, Norway Science & Technology University.

Mr. Deo Prasad, Asia-Pacific President of International Solar Energy Society (ISES) and Professor of Faculty of the Built Environment, University of New South Wales, Sydney, Australia.

Mr. M.Norbert Fisch, Professor of TU Braunschweig and president of the Institute of Architecture and Solar Energy Technology, Germany.

Mr. Peter Luscuere, Professor of Department of Architecture, Delft University of Technology, The Netherlands.

Mr. Mitsuhiro Udagawa, President of ISES-Japan and Professor of Department of Architecture, Kogakuin University.

Mr. Lin Xiande, President of Taiwan Green Building Committee and Professor of Faculty of Architecture of Success University, Taiwan.

Mr. Zhong Jishou, Chief Commissioner of Special Committee of Solar Buildings, CRES and Director of CNERCHS.

Mr. Xi Wenhua, Director-General of Gansu Natural Energy Research Institute; Director-General of UNIDO International Solar Energy Center for Technology Promotion and Transfer; Expert in Sustainable Energy Field from United Nations, International Coordinator.

Mr. Huang Qiuping: Deputy General Architect of Huadong Institute of Architectural Design and Research, Shanghai Modern Design Group.

Mr. Feng Ya, Deputy Chief Engineer of Southwest Architecture Design and Research Institute of China; Deputy Director of Special Committee of Building Thermal and Energy Efficiency, Architectural Society of China.

MEMBERS OF THE ORGANIZING COMMITTEE:

It is composed by competition organizer, operator and sponsor. The administration office is a standing body in Special Committee of Solar Buildings, CRES.

农牧民定居青海低能耗住房项目评分表

评比指标	指标说明	分值
规划布局与建筑设计	规划设计、环境利用、建筑创意、建筑设计等技术，鼓励创新	30
主动太阳能利用技术	通过专门设备收集、转换、传输、利用太阳能的技术，鼓励创新	10
被动太阳能利用技术	通过专门建筑设计与建筑构造利用太阳能的技术，鼓励创新	30
采用的其他技术	其他绿色、低碳、安全、健康技术，鼓励创新	10
技术的可操作性	作品的可实施性，技术的经济性和普适性要求	20

农村住房产业化黄石住宅公园项目评分表

评比指标	指标说明	分值
建筑设计与产业化	环境利用、建筑创意、建筑设计、住宅产业化建造等技术，鼓励创新	30
主动太阳能利用技术	通过专门设备收集、转换、传输、利用太阳能的技术，鼓励创新	10
被动太阳能利用技术	通过专门建筑设计与建筑构造利用太阳能的技术，鼓励创新	30
采用的其他技术	其他绿色、低碳、安全、健康技术，鼓励创新	10
技术的可操作性	作品的可实施性，技术的经济性和普适性要求	20

设计任务书及专业术语：（见资料下载）

附件1：农牧民定居青海低能耗住房项目
附件2：农村住房产业化黄石住宅公园项目
附件3：专业术语

奖项设置及奖励形式：

综合奖：获奖作品建筑设计与所选用太阳能技术具有较强的适配性。
一等奖作品2名　　颁发奖杯、证书及人民币50000元奖金（税前）；
二等奖作品4名　　颁发奖杯、证书及人民币20000元奖金（税前）；
三等奖作品6名　　颁发奖杯、证书及人民币5000元奖金（税前）；
优秀奖作品30名　　颁发证书。

技术专项奖：获奖作品在采用的技术或设计方面具有创新，实用性强。

APPRAISAL METHODS:

1. Organizing Committee will check up eligible entries and confirm shortlist entries.
2. Judging Panel will appraise and select out the awarded works.

APPRAISAL STANDARD:

1. The entries must meet the demands of the Competition Requirement.
2. The entries should embody originality and prospective in order to encourage innovation.
3. The submission works should meet the demands of usable function, green and low-carbon, and health and coziness. The building technology and solar energy technology should have adaptability to each other.
4. The submission works should play the role of reducing building energy consumption by utilization of solar energy technology and have feasibility in the aspect of economy and technology.
5. A percentile score system is adopted for the appraisal as follows:

APPRAISAL INDICATORS:

Subject I: Low Energy-Consumption Housing for Farmers in Qinghai Province

APPRAISAL INDICATOR	EXPLANATION	SCORES
Planning and architecture design	Urban planning design, use of environmental resource, creativity of the design, architectural design. Innovation is encouraged	30
Utilization of active solar energy technology	Use of solar energy after collecting, transforming, and transmitting energy by specific equipments. Innovation is encouraged	10
Utilization of passive solar energy technology	Use of solar energy by specific architecture and construction design. Innovation is encouraged	30
Other technologies	Other technologies such as: green, low carbon, safe and healthy technologies. Innovation is encouraged	10
Operability of the technology	Feasibility, efficiency and popularity of relevant technology and economic demands	20

建筑创意奖：获奖作品在规划及建筑设计方面具有独特创意和先导性。

技术专项奖及建筑创意奖作品名额不限，颁发证书。

作品要求：

1. 建筑设计方面应达到方案设计深度，技术应用方面应有相关的技术图纸和指标。

2. 作品图面、文字表达清楚，数据准确。

3. 作品基本内容包括：

3.1 简要建筑方案设计说明（限200字以内），包括方案构思、太阳能综合应用技术与设计创新等。

3.2 农牧民定居青海低能耗住房项目的竞赛作品需进行竞赛用地范围内的规划设计，总平面图比例为1：500～1：1000（含组团规划及环境设计）；农村住房产业化黄石住宅公园项目的竞赛作品需提供产业化产品和建造技术示意图。

3.3 单体设计：

能充分表达建筑与室内外环境关系的各层平面图、外立面图、剖面图，比例1：50～1：100；

能表现出技术与建筑结合的重点部位、局部详图及节点大样，比例自定；

其他相关的技术图、表等。

3.4 建筑效果表现图1～4个。

3.5 参赛者须将作品文件编排在840mm×590mm的展板区域内（统一采用竖向构图），作品张数应为4或6张。中英文统一使用黑体字。字体大小应符合下列要求：标题字高：25mm；一级标题字高：20mm；二级标题字高：15mm；图名字高：10mm；中文设计说明字高：8mm；英文设计说明字高：6mm；尺寸及标注字高：6mm。文件分辨率100dpi，格式为JPG或PDF文件。

4. 参赛者通过竞赛网页上传功能将作品递交竞赛组委会，入围作品由组委会统一编辑板眉、出图、制作展板。

5. 作品文字要求：除3.1"建筑方案设计说明"采用中英文外，其他为英文；建议使用附件3中提供的专业术语。

参赛要求：

1. 欢迎建筑设计院、高等院校、研究机构、太阳能研发和生产企业等单位，组织相关专业的人员组成竞赛小组参加竞赛。

2. 请参赛人员访问 www.isbdc.cn 或 www.house-china.net/isbdc.cn，按照规定步骤填写注册表，提交后会得到唯一的作品编号。一个作品对应一个注册号。

Subject II: Rural Housing Industrialization in the Residential Park, Huangshi City

APPRAISAL INDICATOR	EXPLANATION	SCORES
Architecture design and industrialization	Environmental considerations, creativity of the design, architectural design, and housing industrialization. Innovation is encouraged	30
Utilization of active solar energy technology	Use of solar energy after collecting, transforming, and transmitting energy by specific equipments. Innovation is encouraged	10
Utilization of passive solar energy technology	Use of solar energy by specific architecture and construction design. Innovation is encouraged	30
Other technologies	Other technologies such as: green, low carbon, safe and healthy technologies. Innovation is encouraged	10
Operability of the technology	Feasibility, efficiency and popularity of relevant technology and economic demands	20

THE TASK OF BUILDING DESIGN AND PROFESSIONAL GLOSSARY (Found in Annex)

Annex 1: Low Energy-Consumption Housing for Farmers in Qinghai Province
Annex 2: Rural Housing Industrialization in the Residential Park, Huangshi City
Annex 3: Professional Glossary

PRIZES:

GENERAL PRIZES:
The awards works of building design and selected solar energy technology must be excellent in adaptability.

First Prize: 2 winners
The Trophy Cup, Certificate and Bonus RMB 50,000 (before tax) will be awarded.

Second Prize: 4 winners
The Trophy Cup, Certificate and Bonus RMB 20,000 (before tax) will be awarded.

Third Prize: 6 winners
The Trophy Cup, Certificate and Bonus RMB 5,000 (before tax) will be awarded.

Honorable Mention Prize: 30 winners
The Certificate will be awarded.

PRIZE FOR TECHNICAL EXCELLENCE WORKS:
Prize works must be innovative with practicability in aspect of technology adopted or design.

提交作品时把注册号标注在每个作品的左上角，字高6mm。注册时间2014年6月1日～2015年1月1日。

3. 参赛人员同意组委会公开刊登、出版、展览、应用其作品。

4. 被编入获奖作品集的作者，应配合组委会，按照出版要求对作品进行相应调整。

注意事项：

1. 参赛作品电子文档须在2015年3月1日前提交组委会，请参赛人员访问www.isbdc.cn 或 www.house-china.net/isbdc.cn，并上传文件，不接受其他递交方式。

2. 作品中不能出现任何与作者信息有关的标记内容，否则将视其为无效作品。

3. 组委会将及时在网上公布入选结果及评比情况，将获奖作品整理出版，并对获奖者予以表彰和奖励。

4. 获奖作品集首次出版后30日内，组委会向获奖作品的创作团队赠样书2册。

5. 竞赛活动消息发布、竞赛问题解答均可登陆竞赛网站查询。

所有权及版权声明：

参赛者提交作品之前，请详细阅读以下条款，充分理解并表示同意。

依据中国国家有关法律法规，凡主动提交作品的"参赛者"或"作者"，主办方认为其已经对所提交的作品版权归属作如下不可撤销声明。

1. 原创声明：

参赛作品是参赛者原创作品，未侵犯任何他人的任何专利、著作权、商标权及其他知识产权；该作品未在报纸、杂志、网站及其他媒体公开发表，未申请专利或进行版权登记，未参加过其他比赛，未以任何形式进入商业渠道。参赛者保证参赛作品终身不以同一作品形式参加其他的设计比赛或转让给他方。否则，主办单位将取消其参赛、入围与获奖资格，收回奖金、奖品并保留追究法律责任的权利。

2. 参赛作品知识产权归属：

为了更广泛推广竞赛成果，所有参赛作品除作者署名权以外的全部著作权归竞赛承办单位及冠名单位所有，包括但不限于以下方式行使著作权：享有对所属竞赛作品方案进行再设计、生产、销售、展示、出版和宣传，享有自行使用、授权他人使用参赛作品用于实地建设之权利。大赛主办方对所有参赛作品拥有展示和宣传等权利。其他任何单位和个人（包括参赛者本人）未经授权不得以任何形式对作品转让、复制、转载、传播、摘编、出版、发行、许可使用等。参赛者同

PRIZE FOR ARCHITECTURAL ORIGINALITY:

Prize works must be originally creative and advanced in planning and building design.

The quota of Prize for Technical Excellence Works and Architectural Originality is open-ended. The Certificate will be awarded. In all cases the Jury's decision will be final.

REQIREMENTS OF THE WORK:

1. The submitted drawing sheets should meet the requirements of scheme design level and should be accompanied with relevant technical drawings and technology data.

2. Drawings and text should be expressed in clear and readable way. Mentioned data should be accurate.

3. The submitted work should include:

3.1 A project description (not exceeding 200 words) including the following facts:
Schematic concept design description;
Integration of solar energy technology;
Innovative design.

3.2 For subject I, participants should provide an urban design within the outline of the site of the competition. Participants will provide a site plan (including urban context / urban design) with the scale of 1：500 or 1：1000.

For subject II, participants should provide drawings for a manufactured product.

3.3 Architectural Design:

Participants will provide floor plans, elevations and sections with the scale of 1：50 or 1：100.

Participants should provide detailed drawings (without limitation of scale) that illustrate the integration of technology in the architectural project, as well as any other relevant elements, such as tables, technical charts and diagrams that adequately communicate the proposal.

3.4 Rendering perspective drawing (1~4).

3.5 Participants should arrange the submission into four or six exhibition panels, each 840 mm×590 mm in size (arranged vertically). Font type should be in boldface. Font height is required as follows: title with word height 25 mm; first subtitle with word height 20 mm; second subtitle: word height 15 mm; figure title: word height 10 mm; design description word height 6 mm; dimensions and labels: 6 mm. File resolution: 100 dpi in JPG or PDF format.

4. Participants should send (upload) a digital version of submission via FTP to the organizing committee, who will compile, print and make exhibition panels for shortlist works.

5. Text requirement: The submission should be in English. Participants should use the words from the Professional Glossary in Appendix 3.

PARTICIPATION REQUIREMENTS:

1. Institutes of architectural design, colleges and universities, research institutions and manufacture enterprises of solar energy are welcome to make competition groups with professionals of architecture, structure and equipment to attend the competition.

意竞赛承办单位及冠名单位在使用参赛作品时将对其作者予以署名，同时对作品将按出版或建设的要求作技术性处理。参赛作品均不退还。

3. 参赛者应对所提交作品的著作权承担责任，凡由于参赛作品而引发的著作权属纠纷均应由作者本人负责。

声明：

1. 参与本次竞赛的活动各方（包括参赛者、评委和组委），即表明已接受上述要求。

2. 本次竞赛的参赛者，须接受评委会的评审决定作为最终竞赛结果。

3. 组委会对竞赛活动具有最终的解释权。

4. 为维护参赛者的合法权益，主办方特提请参赛者对本办法的全部条款、特别是"所有权及版权"声明部分予以充分注意。

附件1：
农牧民定居青海低能耗住房项目

一、农牧民定居青海低能耗住房项目气候条件

农牧民定居青海低能耗住房项目用地位于青海省湟源县西南部日月藏族乡兔尔干村，北纬36°32′，东经101°08′，处于青藏高原东缘，为青海省农业向牧业区过渡的地带，是一个典型的农牧业结合区，海拔从东向西南抬高，全村平均海拔3000m以上。

日月藏族乡兔尔干地区地处内陆，属于大陆性季风气候，具有日照时间长，太阳辐射强，春季干旱多风，夏季短促凉爽，秋季阴湿多雨，冬季漫长干燥等特征；气温日较差大，年较差小，气温随海拔高度的增加而降低，结冻期长，无霜期短。年平均气温约3.0℃，最高气温28.3℃（2010年7月28日），最低气温-23.5℃（2011年1月15日）。4~10月平均降水量为444.6mm。其中，最多年份为533.6mm（2012年），最少年份为384.7mm（2011年）。最长连续降水日数为16天，出现时间在2009年8月15日至29日，降水量为76.7mm。青海全年不供冷，冰冻线1.5m。太阳辐射资源丰富，年日照平均时数为2718.6h。全年以西北风为主，其中冬季主导风向为偏西风，10min极大风速为24.7m/s（2012年12月）；夏季以偏东风为主，极大风速为17.0m/s（2009年6月）；年平均风速为1.73m/s，风向频率以西北风最多。

结合日月藏族乡兔尔干地区的气候特点，竞赛需着重解决冬季集热采暖和夏季通风降温问题。其中，5~9月可不采暖，其他月份均有采暖需求。

2. Please visit www.isbdc.cn or www.house-china.net/isbdc.cn. You may fill the registry according to the instruction and gain an exclusive number of your work after submitting the registry. One work only has one registration number. The number should be indicated in the top left corner of each submission work with word height in 6mm. Registration time: 1st June, 2014 – 1st January, 2015.

3. Participants must agree that the Organizing Committee may publish, print, exhibit and apply their works in public.

4. The authors whose works are edited into the publication should cooperate with the Organizing Committee to adjust their works according to the requirements of press.

IMPORTANT CONSIDERATION:

1. Participant's digital file must be uploaded to the organizing committee's FTP site (www.isbdc.cn or www.house-china.net/isbdc.cn) before 1st March, 2015. Other ways will not be accepted.

2. Any mark, sign or name related to participant's identity should not appear in, on or included with submission files, otherwise the submission will be deemed invalid.

3. The Organizing Committee will publicize the process and result of the appraisal online in a timely manner, compile and publish the awarded works. The winners will be honored and awarded.

4. In 30 days after the collection of works being published, 2 books of award works will be freely presented by the Organizing Committee to the competition teams who are awarded.

5. The information concerning the competition as well as explanation about all activities may be checked and inquired in the website of the competition.

ANNOUNCEMENT ABOUT OWNERSHIP AND COPYRIGHT:

Before submitting the works, participants should carefully read following clauses, fully understand and agree with them.

According to relevant national laws and codes it is made sure by the competition sponsors that all "participants" or "authors" who have submitted their works on their own initiative have received following irrevocable announcement concerning the ownership of their works submitted:

1. Announcement of originality:

The entry work of the participant is original, which does not infringe any patent, copyright, trademark and other intellectual property; it has not been published in any newspapers, periodicals, magazines, webs or other media, has not been applied for any patent or copyright, not been involved in any other competition, and not been put in any commercial channels. The participant should assure that the work has not been put in any other competition by the same work form in its whole life or legally transferred to others, otherwise, the competition sponsors will cancel the qualification of participation, being shortlisted and awarded of the participant, call back the prize and award and reserve the right of legal liability.

2. The ownership of intellectual property of the works:

In order to promote competition results, the participants relinquish copyright of

1. 基本气象资料

气象参数：北纬36°32′、东经101°08′、测量点海拔高度3138m。

月份	空气温度 °C	相对湿度 %	水平面日太阳辐射 kWh/(m²·d)	大气压力 kPa	风速 m/s	土地温度 °C	月采暖度日数 °C·d	供冷度日数 °C·d	日照时数 h
1月	-9.9	42.3	3.35	69.5	1.7	-9.6	846	0	213.2
2月	-7.1	42.5	4.23	69.4	1.7	-5.7	702	0	213.2
3月	-2.6	43.6	5.02	69.5	2.0	-0.3	624	0	239.9
4月	2.5	45	5.80	69.7	2.6	5.7	457	0	239.9
5月	6.8	53.4	5.78	69.8	1.8	10.2	355	4	239.9
6月	10.4	59.5	5.67	69.8	1.5	13.2	234	31	266.5
7月	12.6	62.9	5.80	69.8	1.2	14.8	174	79	239.9
8月	11.4	66.5	5.42	70.0	1.3	13.3	198	59	239.9
9月	7.1	68.4	4.47	70.1	1.6	8.8	317	7	186.6
10月	1.5	62.2	4.01	70.2	1.6	3.0	497	0	213.2
11月	-3.4	43.5	3.49	70.0	2.2	-2.7	623	0	213.2
12月	-7.8	41.7	2.99	69.8	2.7	-7.8	781	0	213.2
年平均数	1.8	52.6	4.67	69.8	1.8	3.6	5808	180	2718.6

2. 采暖通风与空气调节室外气象参数

参数	夏季	冬季
空气调节计算干球温度（°C）	33	18~20
空气调节计算湿球温度（°C）	28	15~18
空气调节计算日均温度（°C）	28~30	20~25
通风计算干球温度（°C）	30~33	17~21
空气调节计算相对湿度（%）	48%	30%
平均风速（m/s）	1.7	1.75
风向	西北偏北	西北偏西

all works to competition administrators and titled unit except authorship. It includes but is not limited to the exercise of copyright as follows: benefit from the right of the works on redesigning, production, selling, exhibition, publishing and publicity; benefit from the right of the works on construction for self use or accrediting to others for use. Without accreditation any organizations and individual (including authors themselves) cannot transfer, copy, reprint, promulgate, extract and edit, publish and admit to use the works by any way. Participants have to agree that competition administrators and titled unit will sign the name of authors when their works are used and the works will be treated for technical processing according to the requirements of publication and construction. All works are not returned to the author.

3. All authors must take responsibility for their copyrights of the works including all disputes of copyright caused by the works.

ANNOUNCEMENT:

1. It implies that everybody who has attended the competition activities including participants, jury members and members of the Organizing Committee has accepted all requirements mentioned above.

2. All participants must accept the appraisal of the jury as the final result of the competition.

3. The Organizing Committee reserves final right to interpret for the competition activities.

4. In order to safeguard the legitimate rights and interests of the participants, the organizers ask participants to fully pay attention to all clauses in this document, especially some clauses with blue colors.

Annex 1：

Low Energy—Consumption Housing for Farmers in Qinghai Province

Climate Condition for Low Energy-Consumption Housing for Farmers in Qinghai Province

The Project of Low Energy-Consumption Housing for Farmers in Qinghai Province is set in the Tibetan village Tuergan in the Riyue, a town in the southeast of Huangyuan county, Qinghai province. Located at N36°32′, E101°08′ on the east rim of Qinghai-Tibetan Plateau, it's a transition belt from agricultural zone to herding zone in Qinghai, a typical combination of both industries. The average altitude is above 3000m, lower on the east side and higher on the southwest side.

Tuergan is an inland village with continental monsoon climate. It has long insolation duration and strong solar radiation, the spring being dry and windy; summer short and cool; autumn humid and rainy; winter long and dry. The diurnal temperature is very high but the annual temperature range is rather low, 28.3°C in the highest (07/28/2010), and -23.5°C in the lowest (01/15/2011), averaging on 3°C annually. The temperature goes lower along with higher altitude. It has long frozen period and short frost-free season. The average precipitation from Apr. to Oct. is 444.6mm, once reaching the higher in 2012, as 533.6mm

二、农牧民定居青海低能耗住房项目设计任务书

1. 项目背景

项目用地位于青海省湟源县西南部日月藏族乡兔尔干村，是日月藏族乡的乡政府所在地。兔尔干村地理位置极为独特，这里是黄土高原的边缘，青藏高原的起点，农业区与牧业区的分界线，季风气候带的分界点，具有不可多得的生物多样性、地质多样性和景观多样性；同时这里也是中原文化与藏族文化的分界点，蕴涵着深厚的历史文化、昆仑文化、宗教文化和民族文化，与军屯、城堡、营盘、茶马互市等一同形成了日月山文化。

兔尔干村现有村民 5000 余人，公共服务配套设施初步形成。

兔儿干新型农村社区项目旨在探索以城乡统筹、城乡一体、集约土地、生态宜居、和谐发展为基本特征的小城镇的发展模式，为青海省推进新型城镇化、城乡一体化进程提供示范。竞赛题目结合新型农村社区农牧民定居的建设需求，充分应用太阳能等可再生能源技术，反映当地建筑风格和生活方式，建设低成本、高性能、绿色、低碳、健康的新型农村住房。

2. 自然条件

兔尔干村处于山坡地上，地势由西北向东南倾斜。平均海拔 3100m，属于拉脊山地貌，周边山多地少，沟壑纵横，地形复杂。

项目用地位于 109 国道西北侧，总占地 5.5hm²，场地地形西北高东南低，高程在 3037.60～3021.00m 之间，东西向高差约 16m。用地西南侧紧邻村级道路，用地内道路东北侧平行村级公路现有 110kV 高压线。

3. 基础设施

基地内基础设施完备，已建有市政自来水、排水、雨水、天然气、供电及通信系统。

4. 竞赛场地

竞赛场地位于项目用地的东南侧，规模约 8000m²，高程在 2027～2022m 之间，东西向高差约 5m。

5. 设计要求

（1）在给定的竞赛用地范围内进行不少于 15 幢单体建筑组成的组团规划设计，体现建筑群体与环境的融合，满足农村生活方式的需求。

（2）进行单体农村住房建筑设计，每户占地面积 150m²（含内院），农村住房为 1～2 层，建筑面积 100～120m²，户型设计以满足日常生活使用为基础，应至少包括如下空间：1 个起居室、3 个卧室、厨房、储藏室、卫生间和辅助用房等。农村住房层高宜为 3m。

（3）新建农村住房外观色彩整体协调、明快，与当地环境相呼应，体现青海藏区建筑特色。

（4）应利用当地日照时间长、太阳辐射强的特点，结合区域建筑特色及严寒

and lowest in 384.7mm in 2011. The longest consecutive precipitation lasted 16 days from Aug. 15 to Aug. 29 in 2009, with a gross amount of 76.7mm. Qinghai doesn't provide a cooling system and the anti-freezing line is as high as 1.5m. It enjoys rich solar radiation resources due to an insolation duration of 2718.6h on average annually. The northwest wind prevails most of the year and more wind comes from the west in winter with an extreme wind speed of 24.7m/s in 10min (2012.12). The summer is more dominated by east wind with an extreme wind speed of 17.0m/s (2009.06). The annual wind speed is 1.73m/s.

Considering the climate condition in Tuergan, participants should focus on the heating issue in winter and ventilation and cooling issue in summer. There's no need for heating from May to Sep.

1. Basic Info

Weather Parameter: N 36°32′, E 101°08′; observation point at 3138m.

Month	Temperature °C	Relative Humidity %	Level Solar Radiation kWh/(m²·d)	Air Pressure kPa	Wind Speed m/s	Ground Temperature °C	Gross Heating Degree °C·d	Gross Cooling Degree °C·d	Radiation Hour h
JAN	-9.9	42.3	3.35	69.5	1.7	-9.6	846	0	213.2
FEB	-7.1	42.5	4.23	69.4	1.7	-5.7	702	0	213.2
MAR	-2.6	43.6	5.02	69.5	2.0	-0.3	624	0	239.9
APR	2.5	45	5.80	69.7	2.6	5.7	457	0	239.9
MAY	6.8	53.4	5.78	69.8	1.8	10.2	355	4	239.9
JUN	10.4	59.5	5.67	69.8	1.5	13.2	234	31	266.5
JUL	12.6	62.9	5.80	69.8	1.2	14.8	174	79	239.9
AUG	11.4	66.5	5.42	70.0	1.3	13.3	198	59	239.9
SEP	7.1	68.4	4.47	70.1	1.6	8.8	317	7	186.6
OCT	1.5	62.2	4.01	70.2	1.6	3.0	497	0	213.2
NOV	-3.4	43.5	3.49	70.0	2.2	-2.7	623	0	213.2
DEC	-7.8	41.7	2.99	69.8	2.7	-7.8	781	0	213.2
YEAR	1.8	52.6	4.67	69.8	1.8	3.6	5808	180	2718.6

2. Outdoor weather parameters for heating, ventilation and air conditioning (AC)

Parameter	Summer	Winter
Dry Bulb Temperature by AC Calculation (°C)	33	18~20
Wet Bulb Temperature by AC Calculation (°C)	28	15~18
Daily Average Temperature by AC Calculation (°C)	28~30	20~25
Dry Bulb Temperature by Ventilation Calculation (°C)	30~33	17~21
Relative Humidity by AC Calculation (%)	48%	30%
Average Wind Speed (m/s)	1.7	1.75
Wind Direction	North by Northwest	West by Northwest

竞赛场地请从下列网址下载:www.isbdc.cn或www.house-china.net/isbdc.cn。
The map used for competition could be downloaded from www.isbdc.cn or www.house-china.net/isbdc.cn.

地区气候特点，分析农村住房使用能耗及应用特点，因地制宜地应用太阳能及其他可再生能源系统,解决冬季集热采暖和夏季通风降温问题,并考虑技术的经济性,能够实际应用和示范推广。

附件2：
农村住房产业化黄石住宅公园项目

一、农村住房产业化黄石住宅公园项目气候条件

农村住房产业化黄石住宅公园项目用地位于湖北省东南部黄石市，长江中游南岸，北纬29°30′～30°15′，东经114°31′～115°30′。黄石市属于亚热带季风气候，四季分明，气候温和、湿润，冬寒期短，全年雨量充沛。年平

Low Energy-Consumption Housing for Farmers in Qinghai Province Task Assignment

1. Background

Located in southwest Huangyuan County, Qinghai province, Tuergan is a Tibetan village and the township government of Riyue is. It bears unique geographic location-the rim of Loess Plateau, the start of Qinghai-Tibetan Plateau, the border between agricultural and herding areas as well as the dividing points between monsoon zones. It enjoys high diversity in biology, geology and landscape view. It also separates Han culture from Tibetan culture with a long history of Kunlun culture, religious culture and folk culture. All these factors form the Riyue Mountain Culture together with station troops, castles, camps, tea-horse trade.

There're 5000 villagers in Tuergan. The village is equipped with basic public service facilities.

The Tuergan New Rural Community project aims to explore a small township developing mode featuring the coordination and integration between urban and rural areas, intensive land use, ecology and harmonious development. It will provides as an example of urban and rural integration, pushing forward modern urbanization in Qinghai. The competition subject, taking this intention into consideration, requires full application of solar power and other renewable energy technologies. The work should reflect local architectural style and life style, building new type of rural housing by low cost, high energy-efficiency and green, low-carbon and non-harmful technology.

2. Natural Condition

Tuergan is located in a hillside fields with a slope terrain going from the northwest to southeast. It's part of the Laji Mountains in Qinghai, and its altitude is 3100m on average. It's surrounded by mountains, ravines and gullies, featuring a complicated landscape.

The competition ground is to the northwest of the national road G109, covering an total area of 5.5 hectares. The ground is higher in the northwest and lower in the southeast, ranging from 3037.60m to 3021.00m. It's adjacent to the rural highway on the southwest side, in parallel to the 110kV high voltage lines in the northeast.

3. Infrastructure

The area is fully equipped with basic facilities including tap water, water drainage, nature gas, electricity supply and communication systems.

4. Competition site

The competition site is located in the southeast side of the project with an area about 8000 square meters from 2027~2022m. The altitude difference is about 5m between the west and the east side.

5. Design Requirements

(1) Participants should design a group of at least 15 housing units within the outline of the site. The design should focus on the integration of the architecture in the context and on the needs of rural lifestyle.

(2) Rural housing unit design:

Each household covers a projected ground surface of 150m² (inner yard

均气温 16.4℃，夏季极端最高气温 39.4℃（1951 年 8 月 8 日），冬季极端最低气温 -18.1℃（1977 年 1 月 30 日），最热月（7 月）平均气温 29.2℃，最冷月（1 月）平均气温 3.9℃。无霜期年平均 264 天，最大冻土深度 10cm，最大积雪深度 30cm。年平均降水量 1382.6mm，年平均降雨日 132 天左右，日最大降水量为 360.4mm（1998 年 7 月 22 日）。全年日照 1699h。境内多东南风，年平均风速为 2.17m/s，极大风速 24.8m/s（2008 年 7 月 6 日）。

结合黄石地区的气候特点，竞赛需着重解决冬季集热采暖和夏季通风降温问题。

1. 基本气象资料

气象参数：北纬 29°30′～30°15′、东经 114°31′～115°30′、测量点海拔高度 166m。

月	空气温度 ℃	相对湿度 %	水平面日太阳辐射 kWh/(m²·d)	大气压力 kPa	风速 m/s	土地温度 ℃	月采暖度日数 ℃·d	供冷度日数 ℃·d
1 月	5.2	69.3	2.54	100.7	3.0	5.3	390	3
2 月	7.0	68.6	2.83	100.5	3.1	7.6	306	12
3 月	10.6	71.3	3.18	100.1	3.0	11.4	229	55
4 月	16.6	72.8	3.99	99.5	2.7	17.8	69	197
5 月	21.1	73.2	4.47	99.1	2.6	22.2	9	341
6 月	24.3	77.9	4.54	98.7	2.5	25.3	0	429
7 月	26.4	81.9	4.87	98.5	2.6	27.2	0	519
8 月	25.6	81.8	4.75	98.7	2.4	26.4	0	495
9 月	22.5	76.3	4.12	99.4	2.7	23.3	0	384
10 月	17.9	69.1	3.21	100.0	2.8	18.7	35	254
11 月	12.4	66.9	2.91	100.5	2.9	12.9	161	98
12 月	7.3	65.7	2.57	100.8	3.0	7.4	322	16
年平均数	16.4	73	3.66	99.7	2.8	17.1	1521	2803

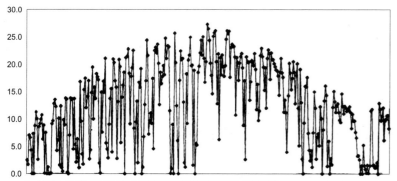

全年每日太阳辐射量折线图　The Line Chart of Daily Solar Radiation in a Year

included). Each rural house should develop on one or two floors, with a total building area of 100~120m². To fulfill the basic needs of daily life, the unit should include the following spaces: 1 living room, 3 bedrooms, a kitchen, a storage room, bathrooms and auxiliary rooms. The suggested floor height is 3m.

(3) New-built rural housing's color should be in harmony with the surroundings, reflecting Qinghai architectural characteristics.

(4) It's suggested to exploit the long sunshine hour and strong solar radiation, and take into consideration local architecture style and local climate condition. Competitors should analyze rural housing energy consumption and other characteristics, and apply solar power and other renewable energy system adjusted to local condition. Competitors should also propose solutions for heating in winter and ventilation and cooling in summer. Economic efficiency of the technology, as well as the feasibility of the design should also be considered.

Annex 2:
Rural Housing Industrialization in the Residential Park, Huangshi City

Climate Condition for The Project of Rural Housing Industrialization in the Residential Park, Huangshi City

The project of Rural Housing Industrialization in the Residential Park is located in Huan gshi city, southeast of Hubei province on the south bank of the Yangtze River at N 29°30′~30°15′, E 114°31′~115°30′. Huangshi is in the subtropical monsoon climate belt, and hence enjoys a moderate and moist climate. The winter is usually short and it rains in the whole year with plentiful rainfall. The annual average temperature is 16.4℃, the extreme minimum temperature is 39.4℃ in summer (08/08/1951) and the extreme minimum temperature is -18.1℃ (01/30/1977). The average temperature of the hottest month, July, is 29.2℃, and the one for the coldest month, January, is 3.9℃. Frost-free period lasts 264 days on year average. The maximum depth of frozen ground is 10cm and the maximum snow depth is 30cm. The average annual precipitation is 1382.6mm, and the average precipitation time is 132 days, with the maximum daily amount occurred in 07/22/1998 as 360.4mm. The annual solar radiation duration is 1699 hours. Southeast wind prevails in the region and the average annual wind speed is 2.17m/s. The extreme wind speed is 24.8m/s (07/06/2008).

Considering the climate condition in Huangshi, participants should focus on the centralized heating issue in winter and ventilation and cooling issue in summer.

1. Basic Meteorological Data

Meteorological Parameters: N 29°30′~30°15′, E 114°31′~115°30′; observation site at 166m.above sea level.

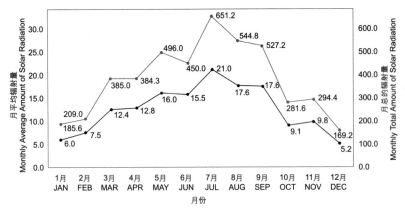

太阳能日辐射量月平均及月总值折线图 The Line Chart of Monthly Average Amount and Monthly Total Amount of Solar Radiation

2. 采暖通风与空气调节室外气象参数

参数	夏季	冬季
空气调节计算干球温度（℃）	35.3	-2.4
空气调节计算湿球温度（℃）	28.4	—
空气调节计算日均温度（℃）	32.2	—
通风计算干球温度（℃）	32.0	0.1
空气调节计算相对湿度（%）	63	72
平均风速（m/s）	2.0	2.6
风向	东南风	东北偏北

二、农村住房产业化黄石住宅公园项目设计任务书

1. 项目背景

竞赛项目位于"黄石住宅公园"中。黄石住宅公园地处中国湖北黄石市"中国住谷"核心区内，为集研究、制造、展示、推广为一体的住宅产业化基地，是黄石市住宅产业化整体布局的对外窗口，旨在通过建设一批适用于夏热冬冷地区的示范住宅，为我国夏热冬冷地区的城镇集合住宅、低层住宅和农村住房提供文化传承与新技术应用的集成示范。

竞赛题目结合黄石住宅公园的定位，重点研究设计适用于夏热冬冷地区的产业化农村住房，解决建筑工业化的设计、建造和应用问题，推广住宅产业化，展示先进适用的住宅设计理念和建筑技术，传播中部地区住宅文化，从而提高生产

Month	Temperature ℃	Relative Humidity %	Level Solar Radiation kWh/(m²·d)	Air Pressure kPa	Wind Speed m/s	Earth Temperature ℃	Monthly Heating Degrees ℃·d	Monthly Cooling Degree ℃·d
JAN	5.2	69.3	2.54	100.7	3.0	5.3	390	3
FEB	7.0	68.6	2.83	100.5	3.1	7.6	306	12
MAR	10.6	71.3	3.18	100.1	3.0	11.4	229	55
APR	16.6	72.8	3.99	99.5	2.7	17.8	69	197
MAY	21.1	73.2	4.47	99.1	2.6	22.2	9	341
JUN	24.3	77.9	4.54	98.7	2.5	25.3	0	429
JUL	26.4	81.9	4.87	98.5	2.6	27.2	0	519
AUG	25.6	81.8	4.75	98.7	2.4	26.4	0	495
SEP	22.5	76.3	4.12	99.4	2.7	23.3	0	384
OCT	17.9	69.1	3.21	100.0	2.8	18.7	35	254
NOV	12.4	66.9	2.91	100.5	2.9	12.9	161	98
DEC	7.3	65.7	2.57	100.8	3.0	7.4	322	16
YEAR	16.4	73	3.66	99.7	2.8	17.1	1521	2803

2. Outdoor Meteorological parameters for heating, ventilation and air conditioning (AC)

Parameter	Summer	Winter
Dry Bulb Temperature by AC Calculation (℃)	35.3	-2.4
Wet Bulb Temperature by AC Calculation (℃)	28.4	—
Daily Average Temperature by AC Calculation (℃)	32.3	—
Dry Bulb Temperature by Ventilation Calculation (℃)	32.0	0.1
Relative Humidity by AC Calculation (%)	63	72
Average Wind Speed (m/s)	2.0	2.6
Wind Direction	Southeast	Northeast by North

Rural Housing Industrialization in the Residential Park, Huangshi City Task Assignment

1. Background

This competition ground is in Huangshi Residential Park in the heart zone of Huangshi, Hubei, which is also known as the Chinese Housing Valley. It's a

效率和农村住房质量。

目前，黄石地区电力资源充沛，钢材、水泥、石膏、砂、石等材料资源丰富，可以建造钢结构、钢混结构、砌块等形式的农村住房。

2. 自然条件

黄石市地跨北纬29°30′~30°15′，东经114°31′~115°30′，位于中国湖北省东南部，长江中游南岸，是武汉经济圈内重要城市。本项目位于黄石西塞山区中国住谷核心区内，紧邻太白湖，占地面积约13.3hm²。本项目将在住宅公园中择地建设，承担农村住房展示、体验、推广等功能。

3. 基础设施

目前住宅公园给水、电市政管网均齐全，无集中燃气管网，无集中排水管网。

4. 竞赛场地

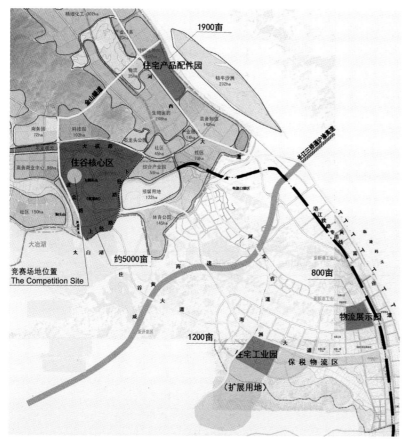

竞赛场地图请从下列网址下载：www.isbdc.cn 或www.house-china.net/isbdc.cn。
Map could be downloaded from www.isbdc.cn or www.house-china.net/isbdc.cn.

housing industrialization base combining research, production, exhibition and publication, and a showcase of the overall deployment of home industrialization in Huangshi. It provides as an example in culture inheritance and technology application for rural and urban collective housing, low-rise dwelling and rural housing.

The subject is related to the location of Huangshi Residential Park. It's a research into industrialized rural housing that can be resided in extremely hot weather in summer and cold weather in winter. It tends to resolve problems including architecture industrialization, construction and application and also popularize housing industry. Participants will show advanced and suitable housing design ideas and technology, spread over central China housing culture and improve productivity and the quality of rural housing.

So far, Huangshi region is rich in electricity, steel, cement, plaster, sand, stone and other resources that could be used to build multiple kinds of rural housing. Steel-structure, reinforced concrete frame structure and building blocks included.

2. Natural Condition

Huangshi is in the southeast of Hubei province, at N 29°30′~30°15′, E 114°31′~115°30′. It's an important city in Wuhan Economic Circle on the south bank of the Yangtze River. This project is located in the core zone of Chinese Housing Valley in Xisai mountain region. It's adjacent to Taibai Lake, covering an area of 133 hectares.

3. Facilities

At present there's water supply system, power grid, internet connection in the Housing Park. There is no central gas pipelines or central sewage pipelines.

4. Competition Ground

5. Design Requirements

(1) Rural housing unit design:

Each household covers a projected ground surface of 150m² (inner yard included). Each rural house should develop on two floors, with a total building area of 200~250m². To fulfill the basic needs of daily life, the unit should include the following spaces: 1 living room, 3 bedrooms, a kitchen, a storage room, bathrooms and auxiliary rooms. The suggested floor height is 3m.

(2) New-built rural housing's color should be in harmony with the surroundings, reflecting Qinghai architectural characteristics.

(3) It's suggested to exploit the local rich solar energy resource, and take into consideration local architecture style and the specific climate condition of the area. Competitors should analyze rural housing energy consumption and other characteristics, and apply solar power and other renewable energy system adjusted to local condition. The participants should focus on the combination of rural housing industry, building method, natural energy technology products and architectural design. Economic efficiency of the technology, as well as the feasibility should also be considered.

(4) Competitors should also propose a pertinent use of active or passive solar energy for heating in winter and ventilation and cooling in summer.

5. 设计要求

（1）典型单体居住建筑设计，每户占地面积150m²，农村住房为二层，以当地农村住房为原型，每户均有自家小院，建筑面积200～250m²，户型设计以满足日常生活使用为基础，应至少包括如下空间：1个起居室、3个卧室、厨房、储藏室、卫生间和辅助用房等。住宅层高宜为3m。

（2）农村住房外观色彩整体协调，与当地环境、特色建筑相呼应，体现本地区地域特征。

（3）应利用当地太阳能资源较丰富的特点，结合建筑设计及夏热冬冷地区气候特点，分析农村住房使用能耗及应用特点，因地制宜地应用太阳能及其他可再生能源系统，探索与太阳能系统结合的农村住房产业化构件和建造方式，重点表达技术、产品与建筑结合的设计方法与建造方式，并考虑技术的经济性，以利于示范推广。

（4）结合建筑方案的特点，通过对太阳能在建筑中的主、被动利用技术的合理选择和应用，着重解决建筑冬季采暖、夏季降温和空气潮湿等问题。

附件3：／Annex 3：
专业术语　Professional Glossary

百叶通风	— shutter ventilation
保温	— thermal insulation
被动太阳能利用	— passive solar energy utilization
敞开系统	— open system
除湿系统	— dehumidification system
储热器	— thermal storage
储水量	— water storage capacity
穿堂风	— through-draught
窗墙面积比	— area ratio of window to wall
次入口	— secondary entrance
导热系数	— thermal conductivity
低能耗	— lower energy consumption
低温热水地板辐射供暖	— low temperature hot water floor radiant heating
地板辐射采暖	— floor panel heating
地面层	— ground layer
额定工作压力	— nominal working pressure
防潮层	— wetproof layer
防冻	— freeze protection
防水层	— waterproof layer
分户热计量	— household-based heat metering
分离式系统	— remote storage system
风速分布	— wind speed distribution
封闭系统	— closed system
辅助热源	— auxiliary thermal source
辅助入口	— accessory entrance
隔热层	— heat insulating layer
隔热窗户	— heat insulation window
跟踪集热器	— tracking collector
光伏发电系统	— photovoltaic system
光伏幕墙	— PV façade
回流系统	— drainback system
回收年限	— payback time
集热器瞬时效率	— instantaneous collector efficiency
集热器阵列	— collector array
集中供暖	— central heating
间接系统	— indirect system
建筑节能率	— building energy saving rate
建筑密度	— building density
建筑面积	— building area
建筑物耗热量指标	— index of building heat loss
节能措施	— energy saving method
节能量	— quantity of energy saving
紧凑式太阳热水器	— close-coupled solar water heater
经济分析	— economic analysis
卷帘外遮阳系统	— roller shutter sun shading system
空气集热器	— air collector
空气质量检测	— air quality test (AQT)
立体绿化	— tridimensional virescence
绿地率	— greening rate
毛细管辐射	— capillary radiation
木工修理室	— repairing room for woodworker
耐用指标	— permanent index
能量储存和回收系统	— energy storage & heat recovery system
平屋面	— plane roof
坡屋面	— sloping roof
强制循环系统	— forced circulation system
热泵供暖	— heat pump heat supply

中文	English	中文	English
热量计量装置	— heat metering device	太阳墙	— solar wall
热稳定性	— thermal stability	填充层	— fill up layer
热效率曲线	— thermal efficiency curve	通风模拟	— ventilation simulation
热压	— thermal pressure	外窗隔热系统	— external windows insulation system
人工湿地效应	— artificial marsh effect	温差控制器	— differential temperature controller
日照标准	— insolation standard	屋顶植被	— roof planting
容积率	— floor area ratio	屋面隔热系统	— roof insulation system
三联供	— triple co-generation	相变材料	— phase change material (PCM)
设计使用年限	— design working life	相变太阳能系统	— phase change solar system
使用面积	— usable area	相变蓄热	— phase change thermal storage
室内舒适度	— indoor comfort level	蓄热特性	— thermal storage characteristic
双层幕墙	— double façade building	雨水收集	— rain water collection
太阳方位角	— solar azimuth	运动场地	— schoolyard
太阳房	— solar house	遮阳系数	— sunshading coefficient
太阳辐射热	— solar radiant heat	直接系统	— direct system
太阳辐射热吸收系数	— absorptance for solar radiation	值班室	— duty room
太阳高度角	— solar altitude	智能建筑控制系统	— building intelligent control system
太阳能保证率	— solar fraction	中庭采光	— atrium lighting
太阳能带辅助热源系统	— solar plus supplementary system	主入口	— main entrance
太阳能电池	— solar cell	贮热水箱	— heat storage tank
太阳能集热器	— solar collector	准备室	— preparation room
太阳能驱动吸附式制冷	— solar driven desiccant evaporative cooling	准稳态	— quasi-steady state
太阳能驱动吸收式制冷	— solar driven absorption cooling	自然通风	— natural ventilation
太阳能热水器	— solar water heating	自然循环系统	— natural circulation system
太阳能烟囱	— solar chimney	自行车棚	— bike parking
太阳能预热系统	— solar preheat system		